STORAGE BATTERIES SIMPLIFIED

OPERATING PRINCIPLES — CARE AND INDUSTRIAL APPLICATIONS

A COMPLETE, NON-TECHNICAL BUT AUTHORITATIVE TREATISE DISCUSSING THE DEVELOPMENT OF THE MODERN STORAGE BATTERY, OUTLINING THE BASIC OPERATION OF THE LEADING TYPES

ALSO THE METHODS OF CONSTRUCTION, CHARGING, MAINTENANCE AND REPAIR. ALL PRACTICAL APPLICATIONS OF COMMERCIAL BATTERIES ARE SHOWN AND DESCRIBED

BY

VICTOR W. PAGÉ, M. S. A. E.

AUTHOR OF "AUTOMOBILE STARTING, LIGHTING AND IGNITION SYSTEMS," ETC.

INCLUDES SPECIAL INSTRUCTIONS FOR CARE AND REPAIR OF AUTOMOBILE BATTERIES AND GLOSSARY OF TERMS

THIS BOOK HAS BEEN WRITTEN WITH THE CO-OPERATION OF THE LEADING AMERICAN STORAGE BATTERY MAKERS

Thoroughly Illustrated with Special Charts, Diagrams and Photographs

WARNING

Remember that the materials and methods described here are from another era. Workers were less safety conscious then, and some methods may be downright dangerous. Be careful! Use good solid judgement in your work, and think ahead. Lindsay Publications, Inc. has not tested these methods and materials and does not endorse them. Our job is merely to pass along to you information from another era. Safety is your responsibility.

Write for a catalog or other unusual books available from:

    Lindsay Publications, Inc.
    PO Box 12
    Bradley, IL 60915-0012

STORAGE BATTERIES
Simplified

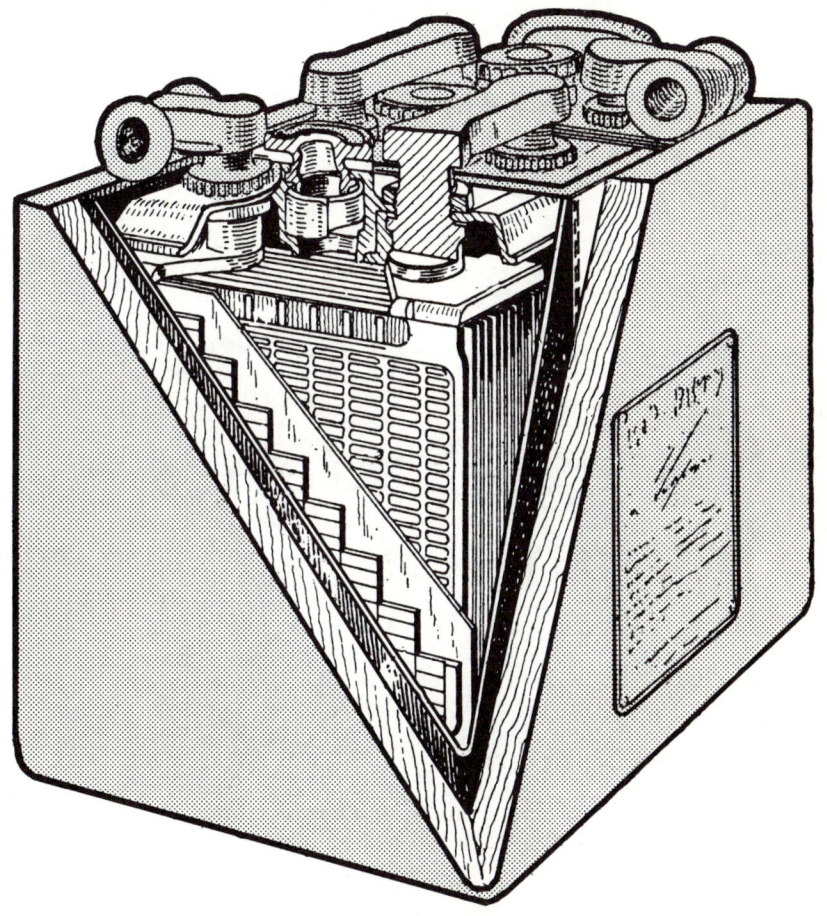

Victor W. Pagé
Lost Technology Series
Reprinted by Lindsay Publications Inc.

Storage Batteries Simplified

Victor W. Pagé, M.S.A.E.

Copyright 1986 by Lindsay Publications, Inc., Bradley, IL 60915. Original copyright 1917 by The Norman W. Henley Publishing Co., published by The Norman W. Henley Publishing Co., New York, 1917.

All rights reserved. No part of this book may be reproduced in any form or by any means without written permission from the publisher.

ISBN 0-917914-47-3

1 2 3 4 5 6 7 8 9 0

INTRODUCTION

THE greatly increasing field of usefulness for the storage battery and the improvements that have been made in its construction are but little appreciated by the average electrician and mechanic. Its use as a source of power for electric automobiles is, of course, well known, but one seldom considers the enormous number of batteries of this type used on the gasoline automobiles. The storage battery is one of the most important parts of the electrical system which includes engine starting and ignition, as well as car lighting of practically every gasoline automobile now manufactured. It is evident that a practical working knowledge of the operating principles and care of the various types of storage batteries now available is desirable. The industrial uses of storage batteries are also increasing, so the student who desires to keep well posted on current engineering development will find much of interest in this treatise. The writer wishes to acknowledge his indebtedness to the following storage battery makers for valuable information and permission to reproduce extracts and illustrations from their literature: Gould Storage Battery Company, Electric Storage Battery Company, Willard Storage Battery Company, U. S. Light and Heat Corporation, and Edison Storage Battery Company. Naturally such co-operation makes a more authoritative and valuable treatise than would otherwise be available. As the experience of these leading authorities on all phases of storage battery engineering is condensed and digested, all the information that the average reader needs may be obtained without consulting a mass of literature. Emphasis is placed on the fact that this work is not intended as a technical treatise but a practical discussion of storage battery operating principles, their maintenance, charging, and chief industrial applications.

THE AUTHOR.

April, 1917.

TABLE OF CONTENTS

CHAPTER I
PAGES

Discovery of Reversible Chemical Action—Chemical Producers of Electricity—Action of Simple Primary Battery—The Dry Cell—Methods of Connecting Dry Cells—Secondary Batteries—Simple Lead Plate Type—Planté, or Formed Plates—Faure, or Pasted Plates—Edison Non-acid Battery Action 13–28

CHAPTER II

How Batteries Differ in Construction—Batteries Using Other Than Lead Plates—Details of Planté Process—Advantages of the Faure Process—Lead Plate Construction—Gould Plates—Exide Plates—Exide "Iron Clad" Plate—Edison Alkaline Battery—Function of Separator 29–52

CHAPTER III

Storage Battery Defects—Sediment in Cells—Sulphation, Cause and Cure—Cadmium Readings—Making Electrolyte—Features of Edison Cells—Tools and Supplies—Taking Down Exide Battery—Taking Down Gould Sealed Cells—Disassembling Willard Battery—Lead-Burning Apparatus—Lead-Burning Process Outlined—Battery Defects and Restoration Summarized................. 53–90

CHAPTER IV

Battery Charging Methods—Currents and Voltages—Electrolytic Rectifiers—Vibrator Rectifiers—Mercury Arc Rectifier—Rotary Converters—Rheostats—Lamp Bank Resistance—Charging Precautions—Charging Vehicle Batteries—Winter Care of Automobile Storage Batteries 91–121

CHAPTER V

Uses of Storage Batteries—Internal Combustion Engine Ignition—Automobile Starting and Lighting Systems—Shifting Gears—Electric Pleasure and Commercial Automobiles—Isolated Lighting Plants—Train Lighting—Mine Locomotives and Street Cars—Electric Launches—Submarine Boats—Miscellaneous Marine Applications—Railway Switch and Signal Service—Stand-by and Booster Service—Drawbridge Operation—Edison Mine Lamp Battery 122–189

CHAPTER VI

Glossary of Storage Battery Terms........................... 190–201

CHAPTER I

Chemical Producers of Electricity—Action of Simple Primary Battery—The Dry Cell—Methods of Connecting Dry Cells—Discovery of Reversible Chemical Action—Secondary Batteries—Simple Lead Plate Type—Planté, or Formed Plates—Faure, or Pasted Plates—Edison Non-acid Battery Action.

Current Production by Chemical Action.—The earliest known method of continuous current generation is by various forms of chemical current producers, which may be either primary or secondary in character. A simple form of cell is shown in section at Fig. 1, A, and as the action of all devices of this character is based on the same principles, it will be well to consider the method of producing electricity by the chemical action of a fluid upon a metal. The simple cell shown consists of a container which is filled with an electrolyte, which may be either an alkaline or acid solution. Immersed in the liquid are two plates of metal, one being of copper, the other zinc. A wire is attached to each plate by means of suitable screw terminals.

If the ends of the plates which are not immersed in the solution are joined together a chemical action will take place between the electrolyte and the zinc plate; in fact, any form of cell consists of dissimilar elements which are capable of conducting electricity immersed in a liquid, which will act on one of them more than the other. The chemical action of electrolyte on the zinc liberates gas bubbles, which are charged with electricity and which deposit themselves on the copper plate. The copper element serves merely as a collecting member, and is termed the "negative" plate, while the zinc which is acted upon by the solution is termed the "positive" member. The flow of current is from the zinc to the copper plate through the electrolyte, and it is returned from the copper plate to the zinc element by the wiring which comprises the external circuit. The terminal on the copper plate is known as the "positive," that on the zinc is called the "negative."

While in the cell shown zinc and copper are used, any other combination of metals between which there exists a difference in electrical condition when one of them is acted upon by a salt or acid may be employed. Any salt or acid solution will act as an electrolyte if it will combine chemically with one of the ele-

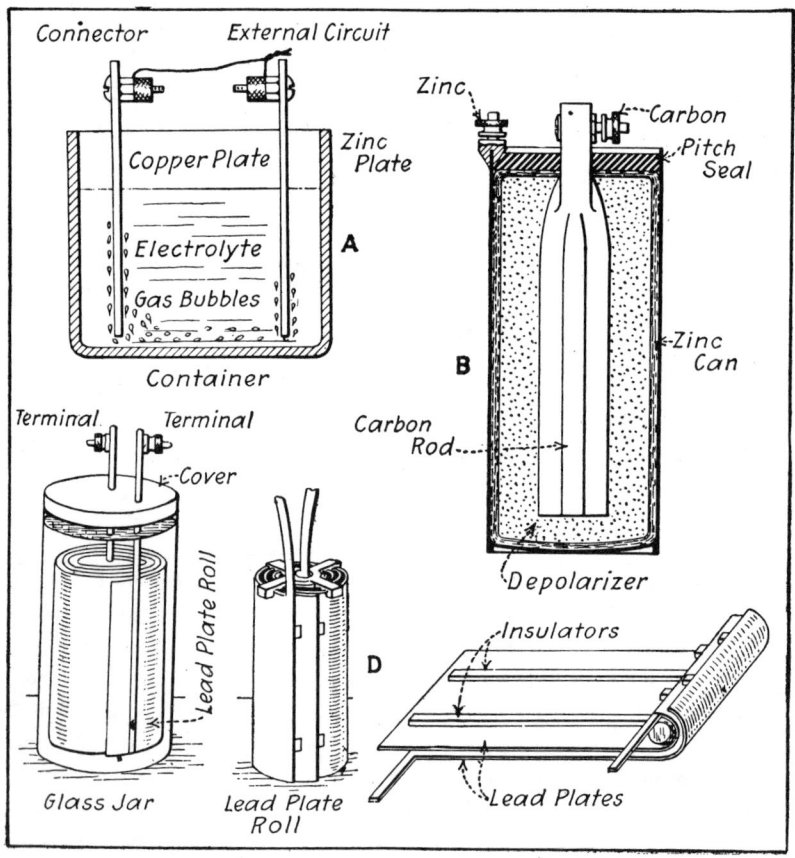

Fig. 1.—Types of Chemical Producers. A—Elementary Primary Cell. B—Construction of Dry Cell. C—Simple Form of Secondary Battery. D—The Lead Plate Roll and How It is Made.

PRODUCTION BY CHEMICAL ACTION 15

ments and if it does not at the same time offer too great a resistance to the passage of the electric current. The current strength will vary with the nature of the elements used, and will have a higher value when the chemical action is more pronounced between the positive member and the electrolyte.

As the vibrations which obtain when the automobile is driven over highways makes it difficult to use primary cells in which there is a surplus of liquid, a form of cell has been devised in which the liquid electrolyte is replaced by a solid substance which cannot splash out of the container even if the cell is not carefully sealed. A current producer of this nature is depicted in section at Fig. 1, B. This is known as a dry cell, and consists of a zinc can, in the center of which a carbon rod is placed. The electrolyte is held close to the zinc or negative member by an absorbent lining of blotting paper, and the carbon rod is surrounded by some depolarizing material. The top of the cell is sealed with pitch to prevent loss of depolarizer.

The depolarizer is needed that the cell may continue to generate current. When the circuit of a simple cell is completed the current generation is brisker than after the cell has been producing electricity for a time. When the cell has been in action the positive element becomes covered with bubbles of hydrogen gas, which is a poor conductor of electricity and tends to decrease the current output of the cell. To prevent these bubbles from interfering with current generation some means must be provided for disposing of the gas. In dry cells the hydrogen gas that causes polarization is combined with oxygen gas evolved by the depolarizing medium, and the combination of these two gases produces water, which does not interfere with the action of the cell. Carbon is used in a dry cell instead of copper because it is a cheaper material, and the electrolyte is a mixture of sal ammoniac and chloride of zinc, which is held in intimate contact with the zinc shell, which forms the negative element by the blotting-paper lining.

When it is desired to obtain more amperage or current quantity than could be obtained from a single cell, they are joined in series-multiple connection. With this method of wiring two or more sets of four cells which have been joined in series are used. The zinc of one set is joined with the zinc of the other, and the two carbons are similarly connected. Any number of sets may be connected in series multiple, and the amperage of the combination is increased proportionately to the number of sets joined together in this manner.

When dry cells are connected in series, the voltage of one

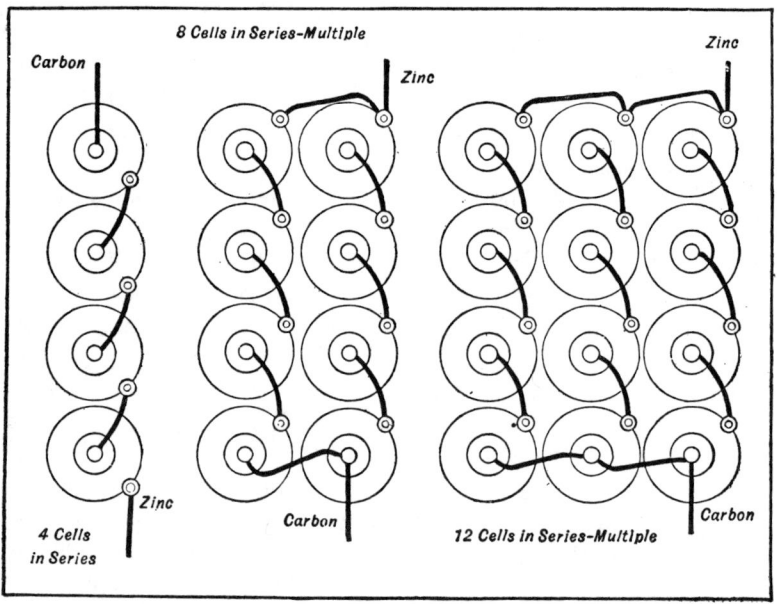

Fig. 2.—Methods of Joining Dry Cells to Form Batteries of Varying Value.

cell is multiplied by the number of cells, and the amperage obtained from the set is equal to that of one cell. When connected in series multiple, as shown at Fig. 2, the amperage is equal to two cells, and the voltage produced is equivalent to that obtained from four cells. When twelve cells are joined in series multiple, the amperage is equal to that of one cell multiplied by three, while the voltage or current pressure is equal to that produced by one cell multiplied by the number of cells which are in series in any one set. By properly combining dry cells in this manner batteries of any desired current strength may be obtained.

The terms "volt" and "ampere" are merely units by which current strength is gauged. The "volt" is the unit of pressure or potential which exists between the terminals of a circuit. The "ampere" measures current quantity or flow, and is independent of the pressure. One may have a current of high amperage at low potential or one having great pressure and but little amperage or current strength. Voltage is necessary to overcome resistance, while the amperage available determines the heating value of the current. As the resistance to current flow increases, the voltage must be augmented proportionately to overcome it. A current having the strength of one ampere with a pressure of one volt is said to have a value of one watt, which is the unit by which the capacity of generators and the amount of current consumption is gauged.

One of the disadvantages of primary cells, as those types which utilize zinc as an active element are called, is that the chemical action produces deterioration and waste of material by oxidization. Dry cells are usually proportioned so that the electrolyte and depolarizing materials become weaker as the zinc is used, and when a dry cell is exhausted it is not profitable to attempt to recharge it because new ones can be obtained at a lower cost than the expense of renewing the worn elements would be.

Principles of Storage Battery Construction.—Some voltaic couples are reversible, i. e., they may be recharged when they have become exhausted by passing a current of electricity through them in a direction opposite to that in which the current flows on discharge. Such batteries are known as "accumulators" or "storage batteries." A storage battery belies its name, as it does not really store current, and its action is somewhat similar to that of the simpler chemical cell previously described. In its simplest form a storage cell would consist of two elements and an electrolyte, as outlined at Fig. 1, D. The storage battery differs from the primary cell in that the elements are composed of the same metal before charging takes place, usually lead, instead of being zinc or carbon. One of the plates is termed the "positive," and may be distinguished from the other because it is brown or chocolate in color after charging, while the negative plate is usually a light gray or leaden color. The active material of a charged storage battery is not metallic lead, but oxides of that material.

The simple form shown at Fig. 1, C, consists of two plates of lead, which are rolled together, separated by insulating bands of rubber at the top and bottom to keep them from touching. This roll is immersed in an electrolyte composed of a weak solution of sulphuric acid in water. Before such a cell can be used it must be charged, which consists of passing a current of electricity through it until the lead plates have changed their nature. After the charging process is complete the lead plates have become so changed in nature that they may be considered as different substances, and a chemical action results between the oxidized plate and the electrolyte and produces current just as in the simple cell shown at Fig. 1, A. When the cell is exhausted the plates return to their discharged condition and are practically the same, and as there is but little difference in electrical condition existing between them, they do not deliver any current until electricity has been passed through the cell so as to change the sulphate on the positive lead plates to peroxide of lead.

The changes that occur in the plates of a storage battery when a current is passed through them and the electrolyte in which they are placed results in altering their nature to such an extent that they act just the same as two elements of more widely differing nature, such as zinc and copper or zinc and carbon might. Primary cell action depends on the wasting of one of the elements, and the only way that more current can be obtained when the cell is exhausted is by replenishing the electrolyte and also putting in a new plate of metal in place of that eaten away by the acid.

Discovery of Reversible Chemical Action.—The basic principle on which storage battery action is based has been known for over a century, as a French scientist, Gautherot, while experimenting with the electrolysis or decomposition of water in 1801 by passing an electric current through it discovered that the silver or platinum wires employed as electrodes for this purpose would send a current of electricity back through a circuit when the battery

DISCOVERY OF REVERSIBLE ACTION

that furnished the decomposing current was removed from them. The electrical flow from the wires was in a reverse direction to that passed through by the main battery, and was naturally very weak. Further studies by other scientists, notably De la Rue, Ritter and Grove, led to the development of a gas battery which was really a variation of the apparatus used in the electrolysis of water. In 1834 Farraday made experiments with lead electrodes that resulted in the development of storage battery forms from which those used to-day are patterned. Following the experiments of Sinsteden, who used plates of nickel, silver and lead in a volt-ampmeter in 1854 and obtained reverse currents from that instrument of sufficient power to raise a wire to red heat after it had been used in measuring a source of electricity, Gaston Planté, who was familiar with these experiments, did the experimental work that formed the basis for the practical storage battery of to-day.

The early Planté battery consisted of two sheets of lead, which were separated from each other by canvas and immersed in a sulphuric acid and water electrolyte. After sending a current of electricity through from primary cells of the Bunsen type, a much more powerful current was obtained from the lead plates than either Farraday, with his lead peroxide element, or Sinsteden, with his combination of metals, had secured, and as a result he is generally credited as being the inventor of the storage battery. After considerable experimenting Planté found that it was possible to increase the current output of a secondary battery by repeated charges and discharges. It was learned that the capacity augmented with use, and that the lead plate surfaces were changed into lead sulphate and lead oxide, and that the coatings penetrated deeper into the plate with each added charge and discharge.

Inasmuch as the knowledge was available to men of science even at that early day (1860) that the action of electric current on lead plates would cause a chemical change that would in turn result in a reverse current flow, the only drawback to making the storage battery of practical value was the lack of economical means for charging the plates. The dynamo had not been perfected to the point that it reached a few years later, so the only

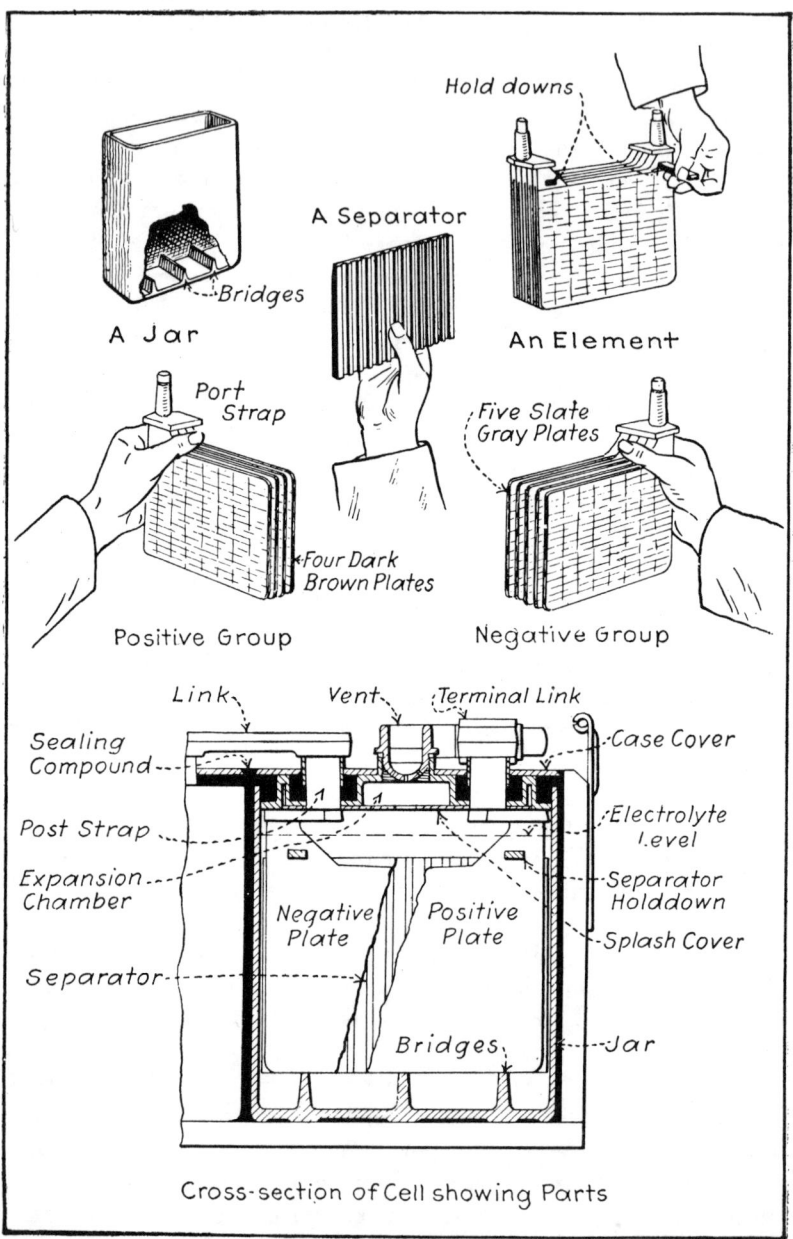

Fig. 3.—Sectional View Showing Typical Storage-Battery Cell and Its Principal Parts.

DISCOVERY OF REVERSIBLE ACTION 21

current means available for "forming" or preparing the lead plates was by the action of expensive primary batteries. The first chemical changes that took place were of very limited depth, and even after long service the chemical action of changed material could hardly be measured. It was only by frequent and repeated charges and discharges, extending over months of time, that it was possible to obtain cells with sufficient capacity to be of practical value. Even though laboring under these disadvantages, Planté was able to make numerous cells of this kind and perform laboratory experiments which created great interest among the scientists of that time.

The chemical change which results in the production of electricity in lead plate batteries is rather complicated, and its exact nature is not definitely known even to-day. It was advanced by Planté that the charging action, or rather chemical change when current was passed through the cell from an outside source, resulted in the formation of lead peroxide (PbO_2) on the positive plate and metallic lead on the negative. Both of these were said to be converted into lead oxide (PbO) when the current was drawn from the battery. Later investigations showed that the formation of lead sulphate also was of enough consequence to be taken into consideration. This reaction on discharge is probably about as follows:

Charged Condition Becomes Discharged Condition.
$$Pb + 2H_2SO_4 + PbO_2 = PbSO_4 + 2H_2O + PbSO_4.$$
—   +   —   +

Therefore, during charging, the plates must be brought to their original state and the sulphate driven out of the plates into the electrolyte, as expressed chemically in the following:

When Discharged.   After Charging.
Positive Plate, $PbSO_4 + O + H_2O = PbO_2 + H_2SO_4.$
Negative Plate, $PbSO_4 + 2H = Pb + H_2SO_4.$

Considering the equations previously outlined, it will be evident that the active material on both plates of a storage battery is changed into lead sulphate when the battery discharges. There are several reasons for considering this theory. The most important is that chemical analysis of a discharged plate has shown large quantities of lead sulphate to exist. The fact that the density of the electrolyte becomes less during the discharge of the cell shows that sulphuric acid is consumed and that water remains. The specific gravity of the electrolyte is greatest when the cell is fully charged. This demonstrates conclusively that during charging the sulphate has been driven out of the plates and into the electrolyte. When a battery is discharged, the sulphate, having been absorbed by the plates, results in a lower specific gravity of the electrolyte. Then, again, considering the matter from an electro-chemist's point of view, it is known that the combination of oxygen and lead as lead oxide would not liberate sufficient electrical energy to account for the voltage of the current produced by the battery on discharge.

Planté, or Formed Plates.—One of the first difficulties met with and one that militated against the development of the practical or commercial type of battery using Planté plates was the great length of time needed and the expensive means of generating the forming current. Later the plates were treated with nitric acid to facilitate the forming action. Other processes have been developed to hasten the formation. In addition to the chemical treatment, which consists of immersing the lead plates in a pickling bath to produce an oxidization before the current acts upon them, there is a mechanical action which will produce the same result and hasten formation. Laminated plates composed of ribbons of lead will form quicker than the solid lead plates, as will elements made up of lead wires or plates where the surface has been grooved with some forming-tool. An electrolytic process consists of making the plate of a lead alloy and eating the foreign matter away to leave a porous lead plate. These processes are described more in detail in the next chapter, which deals specifically with storage battery plate construction.

Faure, or Pasted Plates.—As soon as it was realized that the result of the forming current was the production of lead peroxide on one of the plates, two men, Camille Faure in France and Charles F. Brush in the United States, working independently of each other, devised a process of plate manufacture that materially reduced the cost of construction. Instead of forming the active material by expensive and time-consuming alternating charges and discharges, the common oxides of lead were applied to the surface of the plate in the form of paste, so that the work required of the electric current was reduced appreciably and considerable weight reduction obtained. Litharge, which is rich in lead, was selected for the negative plates, while red lead, which is oxidized more, was used on the plates intended to be positives. The pastes were composed of the oxides mixed with dilute sulphuric acid in the proportion about one part acid to four of water. Such a paste sets very quickly, and only small quantities can be prepared at a time. When the Faure process, as it is called, was first developed, it was believed that the Planté type of plate would be discarded. It was found by practical experience that the new structure developed faults that were not present in the older formation. Pasted plates of early development were found to bend or warp, to enlarge and to shed the active material. In order to eliminate these faults, various ingenious grid patterns were devised.

When storage cells are to be used in automobile work they are combined in a single containing member, as shown at Fig. 4, A, which is a part sectional view of storage battery. The main containing member, a box of wood is divided into three parts by cell jars of hard rubber. Each of these compartments serves to hold the elements comprising one cell. The positive and negative plates are spaced apart by wooden and hard rubber separators, which prevent short circuiting between the plates. After the elements have been put in place in the compartments forming the individual cells of the battery, the top of the jar is sealed by pouring a compound of pitch and rosin, or asphaltum, over cover plates of hard rubber, which keeps the sealing material from running into the cells and on the plates. Vents are provided over each cell, through which gases, produced by charging or discharging, are allowed to escape. These are so formed that while free passage of gas is provided for, it is not possible for the electrolyte to splash out when the vehicle is in motion.

It will be evident that this method of sealing would not be

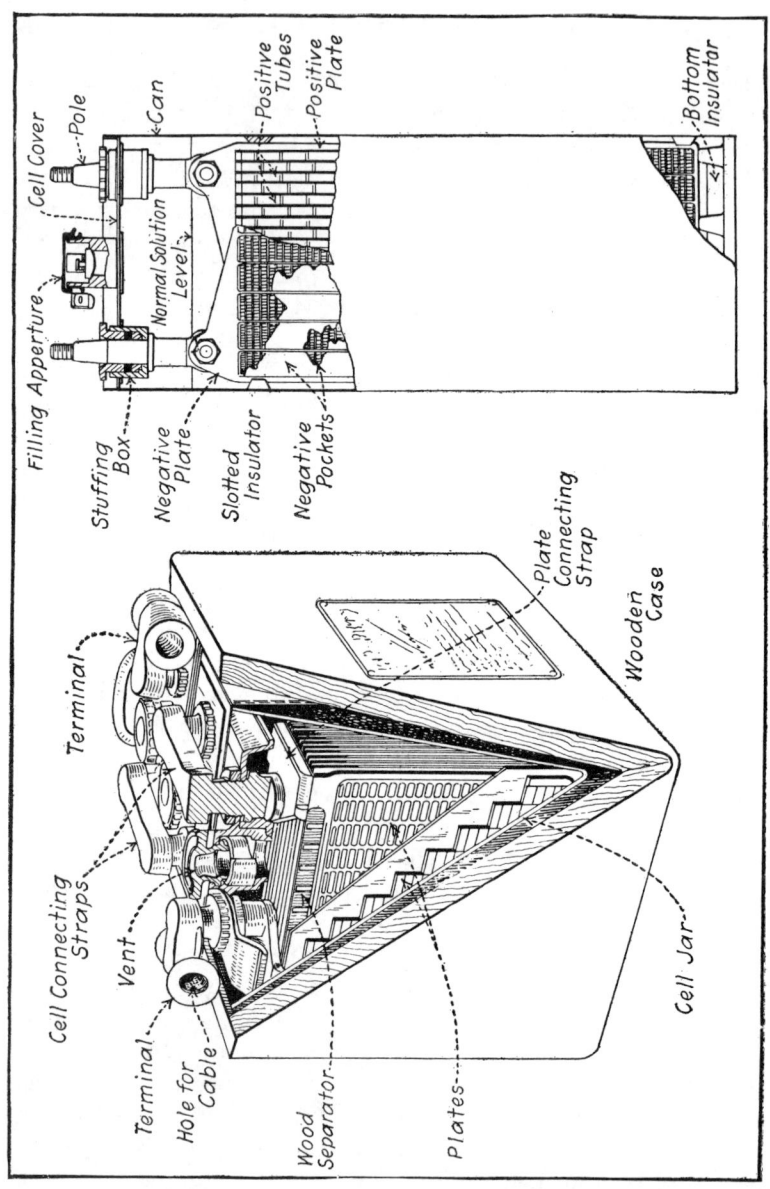

Fig. 4.—Illustrating Standard Batteries Operating on Different Principles. A—Part Sectional View of Automobile Starting, Lighting and Ignition Battery. B—Internal Construction of Edison Alkaline Cell.

practical on a primary cell where the members attacked by the acid had to be replaced from time to time, but in a storage battery only the electrolyte need be renewed. When the plates are discharged they are regenerated by passing a current of electricity through them. New electrolyte can be easily inserted through vents in which caps are screwed. The cells of which a storage battery is composed are nearly always joined together at the factory with bars of lead, which are burned in place, and only two free terminals are provided by which the battery is coupled to the outer circuit.

The capacity of a storage battery depends upon the area and the number of plates per cell, while the potential or voltage is determined by the number of cells joined in series to form the battery. Each cell has a difference of potential of two and two-tenths volts when fully charged, therefore a two-cell battery will deliver a current of four and four-tenths volts, and a three-cell type as shown at Fig. 4, A, will give about six and six-tenths volts between the terminals. While dry cells are often connected in parallel, storage batteries should not be coupled in this manner, as the sets do not divide the load properly unless all the cells are equal in charge, capacity and general condition. A bad cell may so weaken one set that the other set will discharge through it, reducing the charge seriously. When sets are used in parallel it is advisable to test the sets separately. If one set is doing a large proportion of the work, the gravity of the electrolyte will be different in the sets. It is advisable to change parallel sets to series, when practicable, for charging at terminal stations. This insures the same charge being put in each set. For similar reasons all the cells used in a set should be in the same condition and equally well charged to prevent one or more cells reversing in service.

The Edison Storage Battery.—The fundamental principle of the Edison storage battery shown at Fig. 4, B, is the oxidization and reduction of metals in an electrolyte which neither combines with nor dissolves either the active materials or their oxides. Also, an electrolyte which, notwithstanding its decomposition by the action of the electric current, is immediately re-formed in equal quantity, and is, therefore, a practically constant element without change of density or conductivity over long periods of time. A storage battery is commonly looked upon as a receptacle in which to *store electricity*. Electricity is not concrete matter. In fact, nobody knows just *what it is*. Therefore, in the general apprehension of the term, it is not *stored*. Electricity simply causes a chemical change to be effected in certain substances, when it is caused to flow through them. These substances in endeavoring return to their original state, produce electricity.

The following elementary explanation of the action of the non-acid battery is given by the Edison Storage Battery Company, and is so simply expressed and easily understood that it is reproduced in full. If the reader grasps the principles expressed, he will have no difficulty in understanding the chemical action that results in current production.

"Suppose we place two pieces of very thin, bright steel out of doors for a few weeks. They become 'rusted.' The action of the oxygen on the outer layer of the metal has formed it into an oxide commonly known as 'rust.' Now let us place these two pieces of steel in a solution composed of potash and water, and connect them by wires to a small dynamo. The electricity, in flowing from the dynamo through the solution, from one of the plates to the other and back to the dynamo, changes the rust to metallic iron on one of the plates, but causes the other plate to become 'rusted' twice as much as before. Now let us disconnect the plates from the dynamo and connect them, by means of pieces of wire, to an ammeter (an instrument for measuring electricity). Instantly, the excess of oxygen in the rust on the one plate commences to pass back to the bright plate and, by so doing, causes electricity to be generated. Why? Nobody really knows. We have now charged and discharged a primitive storage battery.

"Instead of two thin rusted steel plates, let us mount, say, one hundred such plates, equidistantly spaced, on one rod, and one hundred more on another rod. Now interpose the two groups so the plates of the one group will not touch those of the other and immerse them in a solution of potash. When connected to our dynamo the electricity will flow from one group, through the solution, to the other group, converting the oxide of one group to

THE EDISON STORAGE BATTERY

metallic iron, and increasing the amount of oxide on the other group. We shall be able to get much more electricity from the battery thus formed, because of the greater plate *surface* exposed. We have thus determined that large *surface* is necessary.

"Let us next place a quantity of fine particles of iron rust in two perforated flat steel pockets and, after putting these pockets into potash solution, pass electricity from one to the other, through the solution, as before. All the iron rust in one pocket will be changed to metallic iron, because the oxygen will have passed over to the iron rust in the other pocket, causing this material to possess twice as much oxygen as before.

"Connect the two pockets to your ammeter and you will find that much more electricity is flowing than before, although the two pockets take up much less space than the two hundred steel plates. The reason of this is, the small particles present a very great combined surface to the solution. Suppose, after having made a great number of experiments, you put some iron rust or iron oxide into perforated steel pockets, and mount a number of these pockets in a steel grid or support to form one plate, and place nickel hydrate (a green powder) in perforated steel receptacles, and mount them on another steel grid to form the other plate, then immerse them in a suitable alkaline electrolyte in any kind of container; you have the essential elements of an Edison cell."

The active material of the positive plate of the Edison storage battery is nickel hydrate; that of the negative plate, iron oxide. The electrolyte is a solution of potassium hydrate. The active materials are perfectly insoluble in the electrolyte. When current passes, either on charge or discharge, the electrolyte is broken up into its component parts, which react on the materials with the following results: On charge—Positive oxidized, negative reduced. On discharge—Positive reduced, negative oxidized. The exact chemical changes that go on within the cell are not definitely known, but those occurring during discharge may be approximately represented by the following equations:

Positive: $8K + 6NiO_2 = 2Ni_3O_4 + 4K_2O$.
Negative: $8OH + 3Fe = Fe_3O_4 + 4H_2O$.

The reverse reactions take place on charge. The iron and nickel compounds are probably hydrated, but are here treated as pure oxides, for the sake of simplicity. It will be noted that the same amount of KOH is decomposed, according to the left-hand members of these equations, as is re-formed, simultaneously, as shown on the right. For this reason, the chemical composition, or specific gravity, of the solution does not change appreciably throughout the cycle of charge and discharge.

CHAPTER II

How Storage Batteries Differ in Construction—Batteries Using Other Than Lead Plates—Details of Planté Process—Advantages of Faure Process—Lead Plate Construction—Gould Plates—Combination Faure and Planté Batteries—Exide Plates—Exide "Iron Clad" Plate—Edison Alkaline Battery—Function of Separators.

How Storage Batteries Differ in Construction.—It is not reasonable to expect that any one type of storage battery can be applied universally. Naturally, batteries must be proportioned for the work they are to do. Batteries for stationary work in power plants can be of the open-cell type; those intended for vehicle or boat use should be sealed to prevent danger of splashing out the electrolyte. Various forms of storage batteries are illustrated at Fig. 5. The stationary type in glass or rubber jars is used for isolated lighting plants, telephone, telegraph and signal service and yacht lighting. The larger stationary cells are assembled in lead-lined wood tanks, and are used for electric railways, stand-by service in central stations for lighting and power and in large isolated lighting plants. Special types in rubber jars are assembled in trays for electric vehicle service, and still others for train-lighting service. The usual starting, lighting and ignition battery for automobile use is a unit in which three or more cells are imbedded in insulating compound and carried in a substantial wooden case. The usual practice is to burn all the connecting straps to the plate group terminals and cover the whole with sealing compound. Special gas vents are needed with these batteries. The "couple" type is carried in glass jars and is widely used for signal, fire alarm and private telephone service.

Storage Batteries Using Other Than Lead Plates.—While there is really only one make of cell that is commercially practical that does not use lead plates, this being the Edison, inventors have endeavored to improve storage battery action for some time by trying other combinations of metals. A number of other couples or elements have been found that will permit a reversing chemical

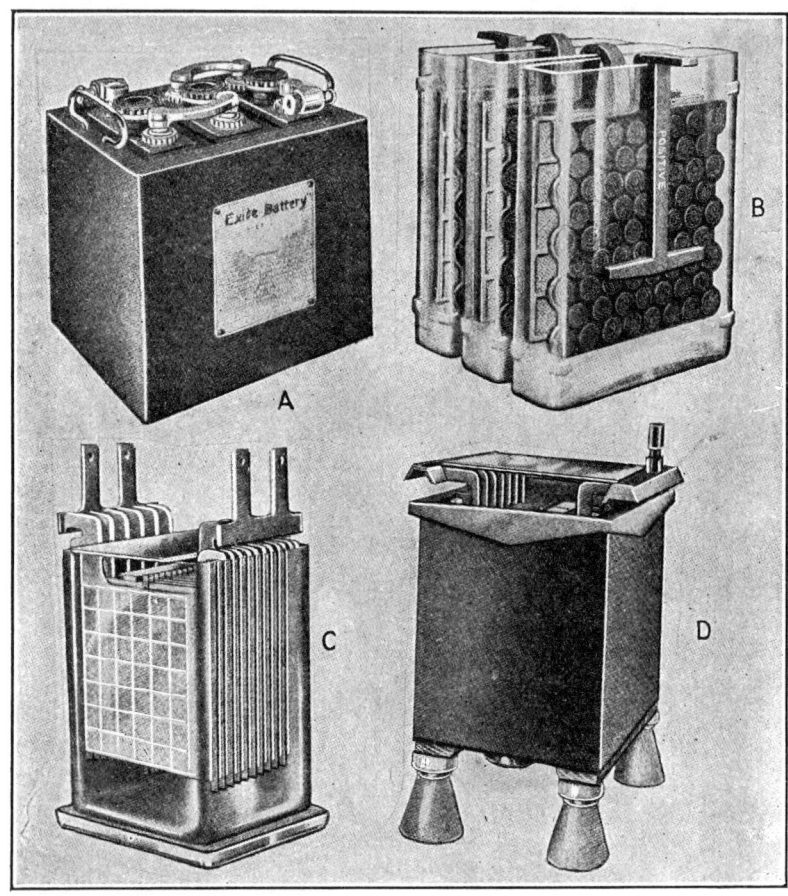

Fig. 5.—How Storage Batteries Differ in Construction. A—Automobile Type. B—Couple Type. C—Glass Jar, Open Type. D—Large Wooden Tank Type for Heavy Duty Service.

action, but most of these are of such a nature that they are interesting additions to a scientist's laboratory rather than contributions to industrial progress. Storage battery plates have been made up

USING OTHER THAN LEAD PLATES

entirely of active material. An experimenter has used a positive plate made entirely of litharge (PbO) mixed with ammonium sulphate $(NH_4)2SO_4$, which is pressed into the desired shape. Chemical treatment converts the plate to lead peroxide. The negative plate is the conventional lead type. This is really a lead plate type, and must be considered as distinct from the non-lead types.

It is a known fact that almost any primary cell can be made to

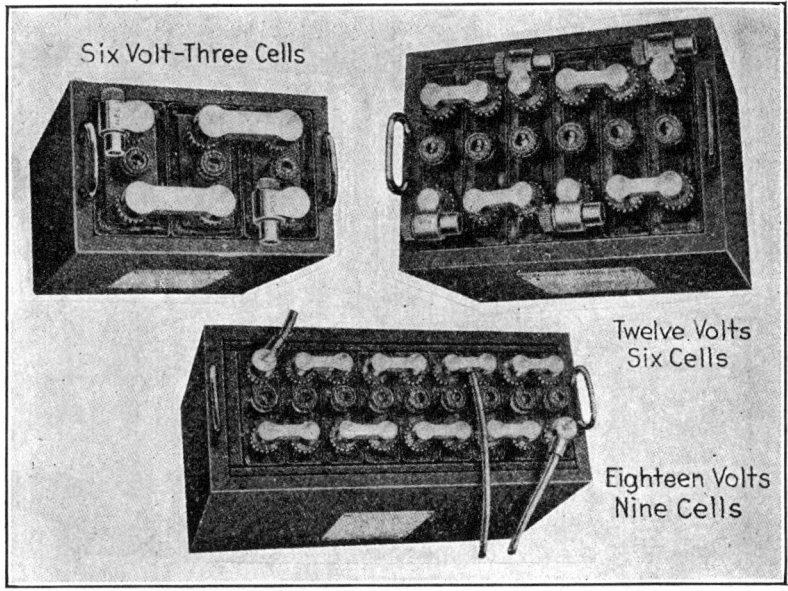

Fig. 6.—Showing Method of Grouping Storage Cells to Form Starting, Lighting and Ignition Batteries.

have some of the characteristics of a storage battery. The experiments in the electrolysis or decomposition of water by using silver or platinum electrodes demonstrates that these substances can be used, though their prohibitive cost renders them only of scientific interest. Zinc has been used for a negative element instead of lead, the surface zinc being converted into zinc sulphate, which dissolved into the electrolyte. Using zinc instead

Fig. 7.—Gould Storage Battery of the Type Used for Isolated Lighting Plants.

USING OTHER THAN LEAD PLATES

of lead gives an increase in voltage, the normal 2.2 augmenting to 2.5 volts. A reduction of weight is also possible, because zinc plates can be lighter than lead one of the same capacity.

The disadvantage is the formation of zinc deposits during charging in the shape of clusters or "trees," which may short-circuit the cell by extending across to the positive element or increase the sediment by dropping to the bottom. Another disadvantage is that the solution will vary in density at different heights. The zinc sulphate is taken from the top of the liquid during charging. Attempts have been made to prevent this by placing the plates horizontally and thus having practically the same density electrolyte surrounding each plate. The fault of this arrangement is that gas bubbles polarize the cell by collecting between the plates. In a vertical plate cell the bubbles rise to the top of the electrolyte and burst, liberating the confined gas, which reaches the atmosphere easily in an open-type cell and through a vent in the sealed types.

Another cell has a negative plate consisting of thin sheet copper amalgamized with zinc. The positives are made of lead leaves perforated with numerous small holes and riveted together, and are formed by the Planté process. The copper-alkali-zinc battery of Lalande and Chaperon and improved by Edison is reversible in action and can be used as a storage battery. When discharged, the positive plate is porous copper; on charging, the decomposition of the electrolyte follows, metallic zinc is deposited on the negative plate, while the porous copper becomes oxidized on the positive. The electrolyte becomes potassium hydrate. This cell may really be considered the ancestor of the modern Edison cell. This storage battery has been used commercially in a limited way, but is not really practical because of its low voltage, that of one cell being but seven-tenths of a volt. That means that three times as many cells would be needed to obtain a certain voltage given by the smaller number of lead plate cells. The weight factor is a serious one that militates against the wide commercial use of low voltage cells. The lead plate type has many practical advantages, but its ability to stand rapid discharge, its great efficiency and its high E.M.F. are among the most important ones.

Fig. 8.—Type of Electric Storage Battery Company Cell for Stand-by and Booster Service.

MAKING PLANTÉ PROCESS PLATES

Making Planté Process Plates.—It is evident that even with the present low cost of electric current makers of storage batteries employing Planté elements must have a more commercial method of forming these than the repeated charge and discharge processes followed by the originator of this type of plate. It is also apparent that the electro-chemical action would form but a very thin layer of active material on plain lead sheets. In order to have a sufficient volume of it to generate an appreciable current it is necessary to provide a larger surface than prevails on an ordinary sheet of lead. To provide more surface, the usual process is to groove the lead plates in order to provide a sufficiently large area so that the forming process will produce an element of sufficient capacity to be commercially practical. The lead peroxide formed on such plates is positively held in the spaces between the ribs or laminations designed to increase the surface.

Various systems of increasing the surface of lead sheets to increase the available area are by grooving, swedging, laminating, scoring or casting the pockets or retaining ribs integrally. A typical Planté process plate used in the Gould battery, with sectional views showing the structure if the plate is cut on lines A-B or C-D, is shown at Fig. 9, and the method of forming clearly outlined at Fig. 10. In the manufacture of these plates the "spinning" process is followed, as this is said to give the greatest possible increased surface and least modification in form of rim and groove. The blank plate, including the lug, is stamped from chemically pure rolled lead and placed in a steel frame, which reciprocates between two revolving mandrels, on which thin circular steel discs and spacing washers are placed. It is on the form and thickness of the discs that the width and shape of the grooves depend, and the thickness of the rib is regulated by the width of the spacing washers. The travel of the frame obviously determines the length of the section to be spun.

The pressure of the "spinning" rolls against the surface of the lead blank is maintained at a uniform point by compressed air, and the ridges and grooves begin to appear as soon as the operation is started. At the start the section is as shown at Fig. 10, B-1. As the spinning discs progress further and further into the lead they displace it and cause it to flow in the form of ribs in the spaces between the cutting discs. The first action is to groove the surface of the lead plate as shown at Fig. 10, B-2, this section becoming more and more like 3 and 4, as the pressure of the disc causes more metal to be displaced. The blank is merely changed in form, as no lead is removed, and there is not cutting and subsequent bending to open up pores in the metal. The discs leave an

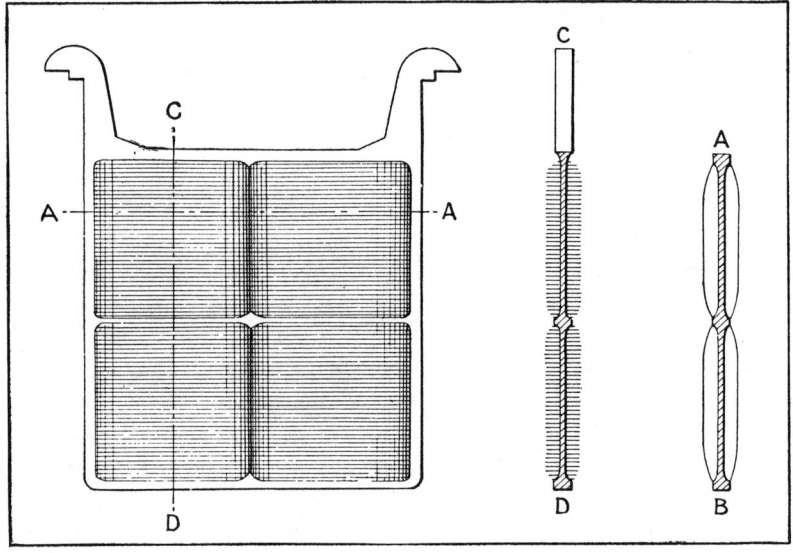

Fig. 9.—Gould Storage-Battery Plate Made by Planté Process.

unspun portion at the end of travel of the frame, in which each individual rib terminates, thus forming a main cross bar at each extremity. The two bars unite at the junction of two spun sections in a single cross bar of diamond section, solid metal, extending the width of the plate. By limiting the depth that the rolls penetrate the blank, it is possible to provide a web of metal that remains as a central conductor and current equalizer.

On the surface of plates thus produced a thin layer of lead peroxide, which is the active material of the positive plate, is formed by the electrolytic process. Negative plates are formed by

MAKING PLANTÉ PROCESS PLATES

a subsequent conversion of the lead peroxide to "spongy" lead, which constitutes the active material for these plates. Finally, the plates are subjected to a special treatment to remove any impurities. An advantage of the spinning process is that the ribs are not cut and subsequently bent, as in some of the other methods of producing Planté plates. It is said that bending, because of the crystalline properties of lead, opens up pores for the penetration of electrolytic action, and that such ribs may be eventually cut off by the chemical action after the battery has been in use

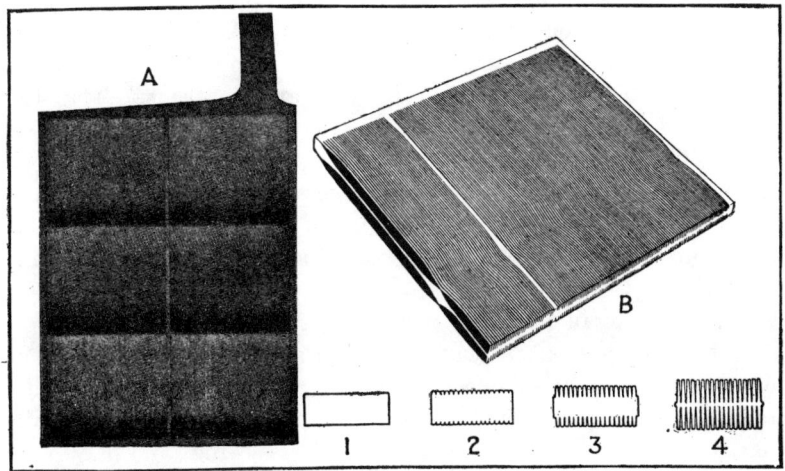

Fig. 10.—Planté Plate and How It is Grooved by Mechanical Means to Facilitate Forming.

for a time. It is also stated that cast lead is not equal to the rolled or spun lead for battery plates on account of its porosity. The parts of a typical Gould cell used for train-lighting service and the method of grouping two of these cells in trays to make for easy handling is clearly shown at Fig. 11.

Manufacture of Faure Type Plates.—Most of the lead plate type storage batteries now used have "pasted" plates instead of the more expensive formed plates. The foundation of a Faure type plate consists of a skeleton or plate grid, such as shown in Figs. 13 and 14, made from an alloy of antimony and lead, to which the active material is mechanically applied. The original Faure cell had both the positive and negative plates coated with red lead. But a comparatively short time was required to change the red lead to lead peroxide on the anode or to metallic lead on the cathode. The great advantage of this construction was the high capacity for unit weight. There is a disadvantage, and that is that if the plates are not very carefully made the active material may drop away from the grid pockets and fall to the bottom of the cell. These have been largely overcome at the present time by forming the grids to hold the applied material more firmly. The reason that pure lead is not used to make grids is that it does not have enough rigidity or strength for use when the active material is applied by mechanical means. The soft lead grids might be bent, which would tend to loosen the active material. This is true to a certain degree of Planté plates, but inasmuch as the active material is generally formed on small surfaces separate from each other, and as it is much thinner than in the applied types, there is not so much danger of the material falling off.

In order to increase the strength of the lead grid it is necessary to add some substance that will make a stiffer skeleton, but of course this material should not change the electrical characteristics of the grid to any extent. Antimony is the material ordinarily added, and the proportions of the resulting alloy may vary from 88% lead and 12% of this metal to 98% lead and but 2% of antimony. It is stated that the usual mixture is 4% antimony and 96% lead. Positive grids should have more lead in their composition than the grids intended to be made into negative plates. While in many cases the active material is applied to the plates by hand, it is advanced that machine-pasted plates make more enduring batteries.

Combination Planté and Faure Types.—The "Chloride" battery, which was manufactured for a time by the Electric Storage Battery Company, is a compromise between the two types. In this cell the positive plate is a Planté type, and the negative follows the Faure principle of construction. Finely divided lead is produced by blowing a stream of air against a stream of molten

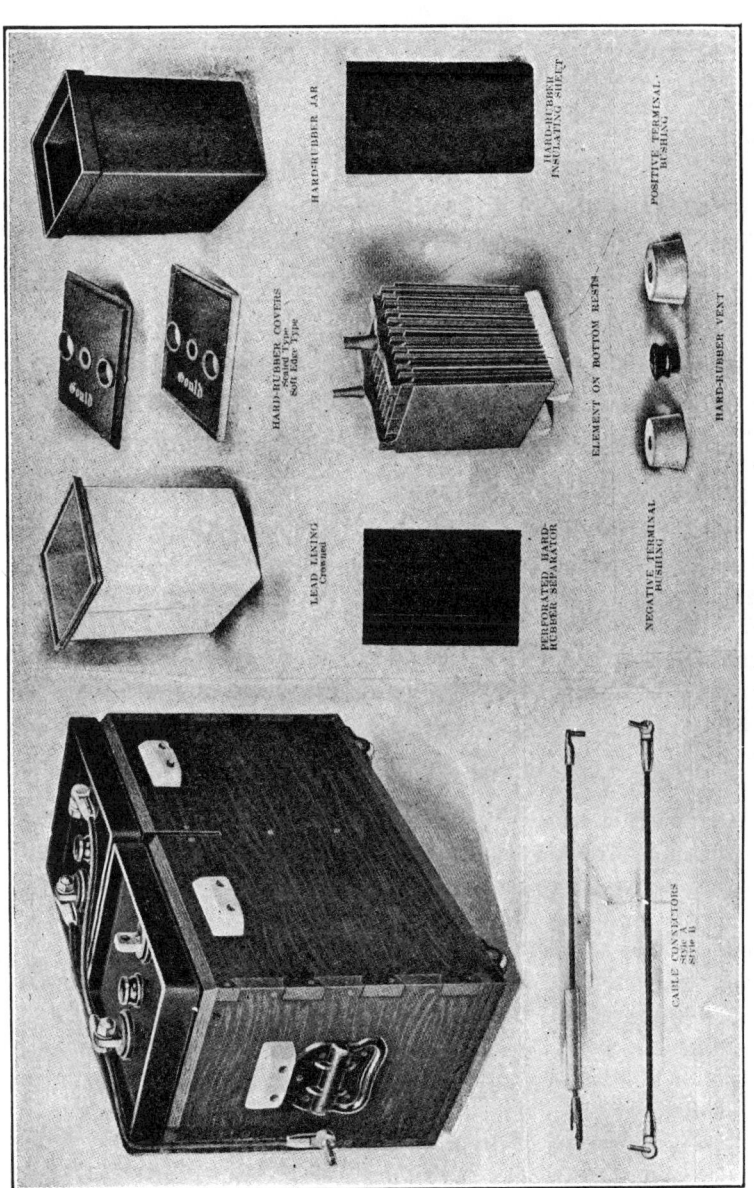

Fig. 11.—The Gould Railway Lighting Battery and Its Principal Parts.

metal, which result in the production of a lead spray, which falls as a powder when cooled. Nitric acid is used to dissolve this powder, which precipitates as lead chloride when hydrochloric acid is added. After this material is washed and dried it forms the basis of the filling of the negative plates. A mixture of this lead chloride and zinc chloride is melted in crucibles and poured into moulds, which produce small tablets about $3/4$ of an inch square and of a thickness varying from $1/4$ to $5/16$ of an inch, depending upon the thickness of the negative plate. These tablets are then assembled in special moulds and held in place by recesses, into which they fit and which prevent movement. They are kept at a distance of about .2 inch from each other and from the mould edges. Molten antimonious lead is then poured in to fill the spaces between the tablets, and to insure a proper flow of metal it is forced into the mould under approximately 75 pounds pressure. Upon cooling, the result is a solid lead grid, in which the small squares of active material are imbedded. The next step is to reduce the lead chloride by placing the plates in a dilute solution of zinc chloride, each plate being separated from its neighbor by a slab of zinc. By assembling the plates in this manner the equivalent of "short circuiting" a cell is obtained, and the lead chloride is reduced to metallic lead. The zinc chloride is removed from the plates by thoroughly washing them.

A new form of negative plate which is now manufactured to replace the chloride type just described consists of a pocketed grid, the openings of which are filled with litharge paste and afterward covered with perforated lead sheets, which are formed by casting integrally with the grid. The grid used for the positive plate is composed of a 5% antimony-lead alloy and is about $7/16$ inch thick, having circular holes about $3/4$ of an inch in diameter, staggered so that the nearest points are about $3/32$ of an inch apart. Close spirals are rolled up of corrugated lead ribbon of the same width as the plate thickness, and these are forced into the circular holes of the plate. The spirals are formed into active material by the electro-chemical process, and during this the spirals expand sufficiently so they fit closer to the grid sides. This form of positive is known as the Manchester plate

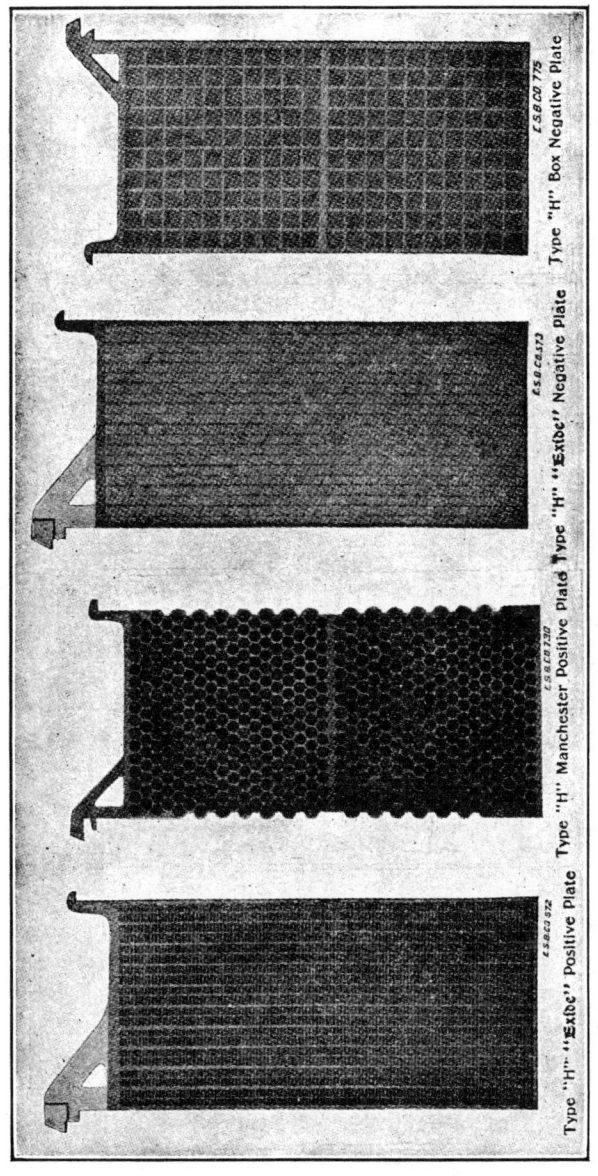

Fig. 12.—Types of Exide Storage-Battery Plates Used in Large Batteries.

and is illustrated at Fig. 12. The box negative plate, the construction of which has been previously described, is also shown at Fig. 12.

"Iron Clad" Exide Battery.—The capacity of the conventional pasted type Exide plate rises in service for a time and then gradually becomes less. The initial rating is conservative, however, so that if a battery is given a proper initial charge it will give its rated discharge at the start. This will gradually increase in use, so that the output becomes greater, and then there is a dropping off from the maximum. This rise in capacity when the battery

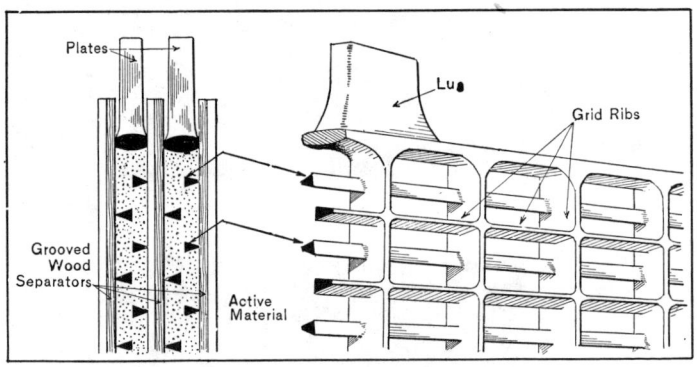

Fig. 13.—Sectional View of Grid, Showing Form of Ribs to Hold Active Material.

is first put into service results from the increasing porosity of the active material on the positive plate. The more porous this active material the better the electrolyte diffuses through it and more lead peroxide is brought into action on each cycle of charge and discharge. This increase in capacity is evidently made at the expense of the positive active material, because as more is brought into action the active mass becomes softer, and the time comes when some of the material must be dislodged when the battery is charged and it will settle to the cell bottom in the form of sediment. This explains why the life of a battery is shortened by too much charging. The capacity will augment just as long as the rate of increase in the porosity of the active material is greater

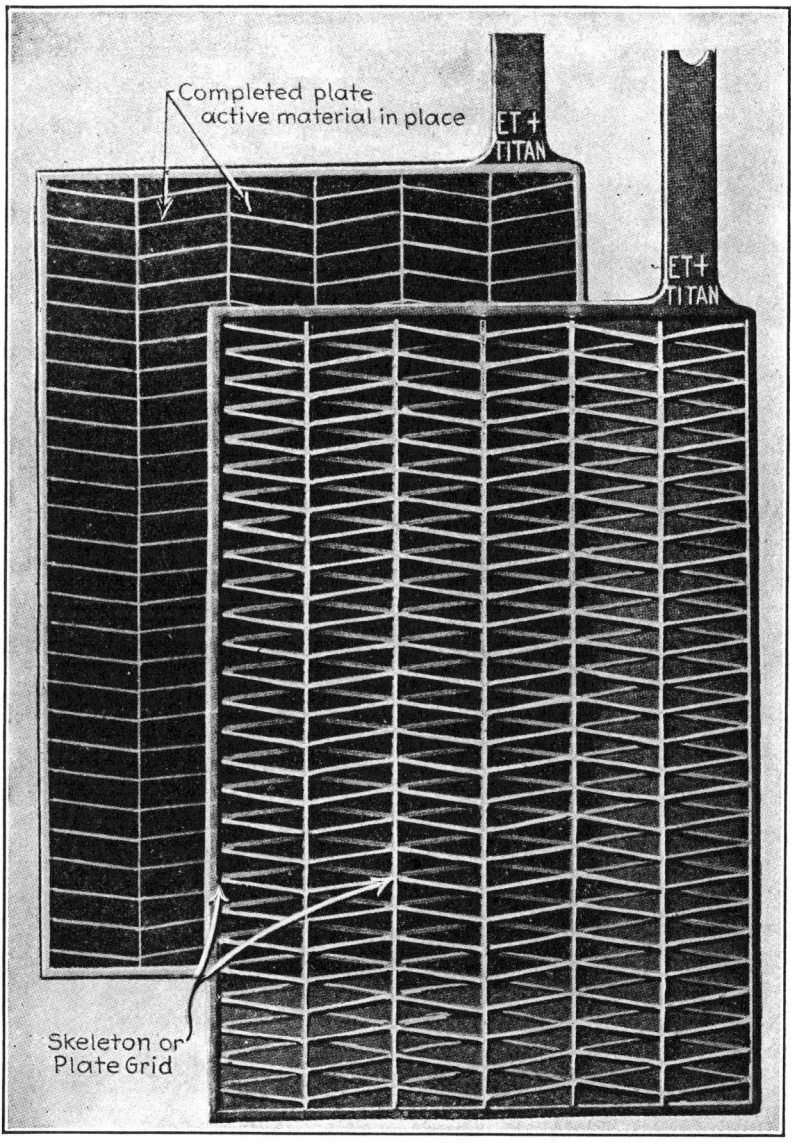

Fig. 14.—The Diamond Grid Plate Before and After Active Material is Put in Place.

than the rate at which the active material loosens from the plate. After a period of use the loss of material will become greater than the gain in porosity, and it is evident that the cell will begin to lose capacity when this condition is reached.

It is evident that if the positive active material could be prevented from dropping off and still be maintained in a healthy operating condition that the plates would have longer life. While improvements have been made from time to time in the construction of the elements, the new form of positive was evolved. This was accomplished by keeping the active material in position by utilizing a pencil of lead peroxide surrounding a conducting core and enclosed in a porous tube having a sufficient elasticity so that as the active material expanded and contracted, because of alterations in its molecular structure, the containing tube compensated for these variations. The positive plate of the "Iron Clad" Exide consists of an alloy framework comprising top and bottom bars integrally connected by conducting cores of the same metal. The uniform pencils of active material surround these cores and are protected by horizontally laminated rubber tubes.

Each tube is formed with narrow vertical ribs diametrically opposite each other, which take the place of the spacing ribs on the ordinary wooden separator and at the same time re-enforce the tube. By thus protecting the active material and holding it in position it remains active for a considerable time. Excellent conductivity and increased accessibility for the electrolyte are obtained, thereby making it possible to secure a relatively high output from a comparatively small quantity of active material. This battery, which is illustrated at Fig. 17 A, having the positive plate shown at B, was given its name because of its remarkable durability. The negative plate of this battery, which is shown at Fig. 17 C, is of the same general construction as the regular Exide negatives, but is made somewhat thicker in order to compensate for the longer life of the positive plate.

The wood separator used between the plates of this battery is a sheet of chemically treated wood and is flat on both sides. No rubber separators are required, inasmuch as the positive plate provides its own separator in having the ribbed rubber tubes to retain

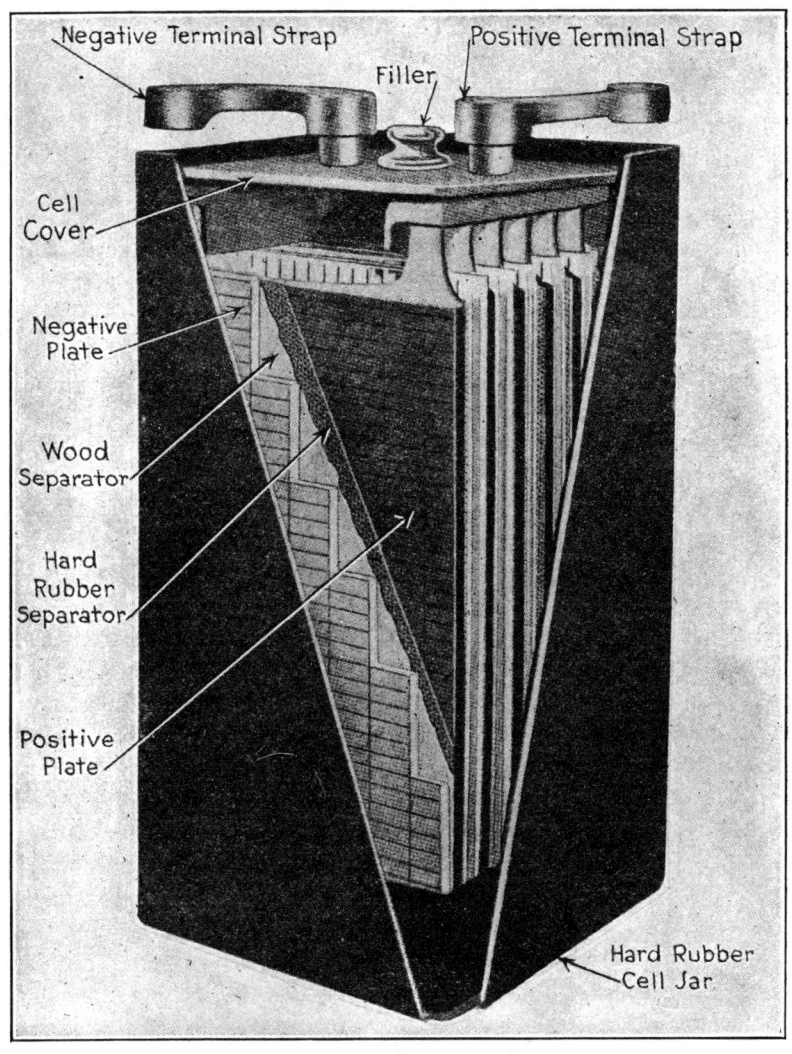

Fig. 15.—Part Sectional View of Type MV 15 Exide Vehicle Battery.

the active material. This is a very popular battery for electric vehicle use because it has a high discharge voltage and is of high efficiency. Flexible pillar strap connectors are regular equipment on "Iron Clad" Exide batteries. These consist of alloy terminals cast around lead plated copper strips, which give greater conductivity and which are more flexible than the stiff pillar-strap connectors used with the Exide standard lead plate batteries. The jar is the same as used for the Exide cells of similar size. A special type of vent is provided, which insures positive retention of the electrolyte yet permits the escape of gas evolved when the battery is charged. The top of the vehicle type cell with vent in place is shown at Fig. 17 E, while the method of sealing is clearly outlined at Fig. 17 D.

The Edison Alkaline Storage Battery.—This is the only battery built of steel. It is the only storage battery having an alkaline solution and using active materials of nickel hydrate (positive) and iron oxide (negative). This construction and principle are said to have important advantages, and some of these are: It is light in weight. It occupies less space. Requires no spare parts. Its steel container is unbreakable. Requires very little attention. It suffers small loss of charge when idle. Does not need frequent hydrometer readings. Its tray assembly and cell connections are simple. It cannot suffer from sulphation or any kindred "disease." Its exclusive use eliminates the need of a battery house. It is not subject to buckling or growing of plates. It may be discharged to zero, or as low as may be desired, without fear of injury. It requires no internal cleaning, the active materials being held securely in perforated steel tubes and pockets. It may be left unused, either charged or discharged, for an indefinite time, without any attention, and suffer no injury. Its cells are hermetically sealed, except for the single filler opening, indicating conclusively that no plate renewals, separator renewals or other repairs are needed or expected. It can be put on charge at any time, regardless of how much or how little of the previous charge has been used; and similarly it may be taken off charge at any time and used, whether fully charged or not.

The positive plates (Fig. 19) consist of a series of perforated steel tubes which are heavily nickel-plated and which are filled with alternate layers of nickel hydroxide and pure metallic nickel in very thin plates. The tube is drawn from a perforated ribbon

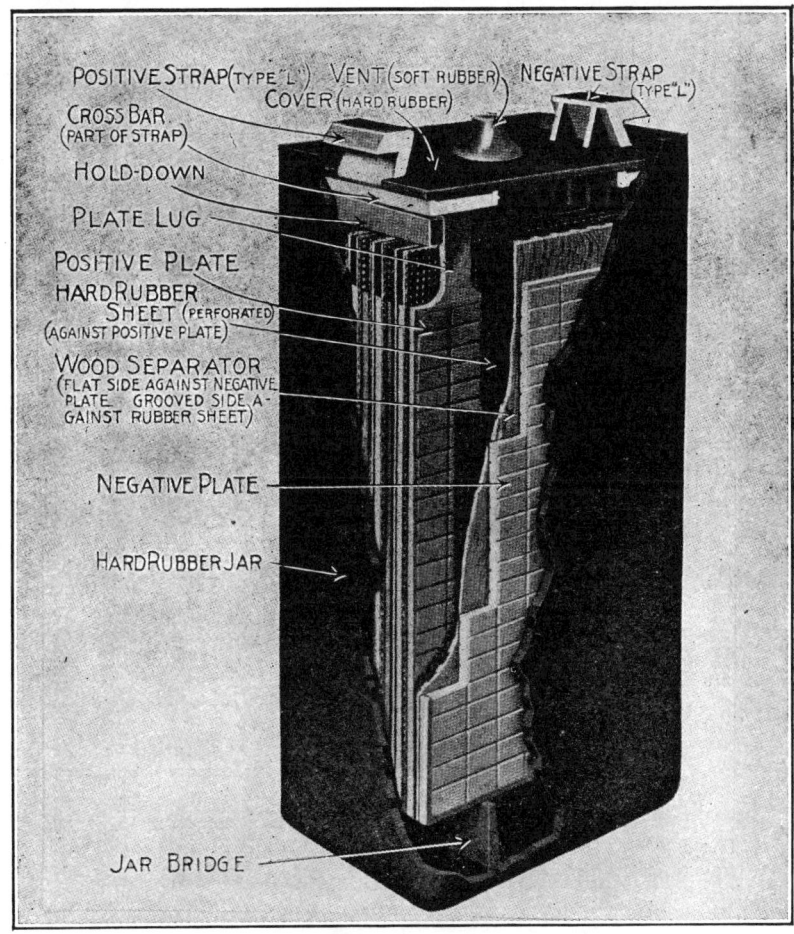

Fig. 16.—Part Sectional View of Gould Pasted Plate Cell.

of steel, nickel-plated, and has a spiral-lapped seam. After being filled with active material it is re-enforced with eight steel bands, which prevent the tube expanding away from and breaking contact with its contents. The negative plate consists of a grid of cold-rolled steel, also heavily nickel-plated, holding a number of rectangular pockets filled with powdered iron oxide. These pockets are also made up of finely perforated steel, nickel-plated. After the pockets are filled they are inserted in the grid and subjected

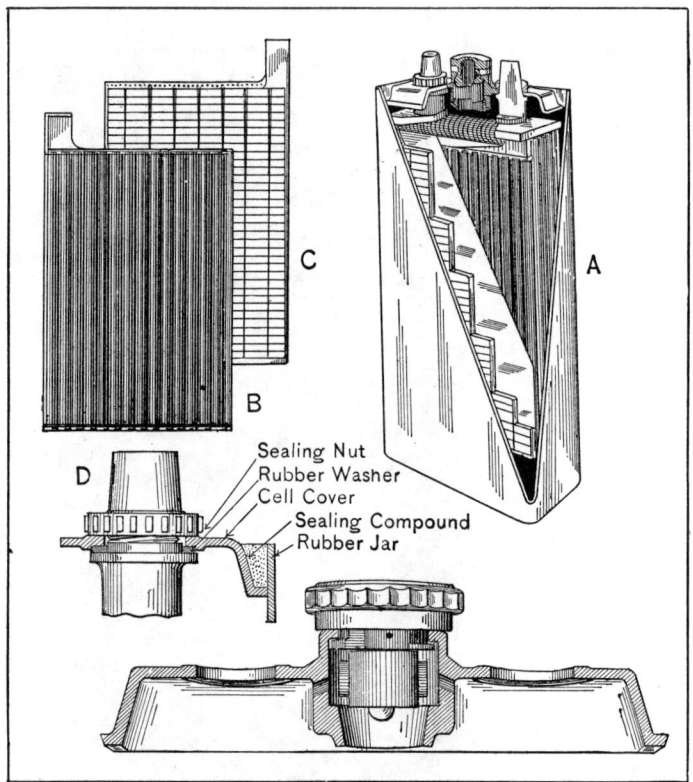

Fig. 17.—Constructional Features of the Exide "Iron Clad" Vehicle Battery.

to considerable pressure between dies, which corrugate the surfaces of the pockets and force them into positive contact with the grids.

These elements are housed in a jar or container made from

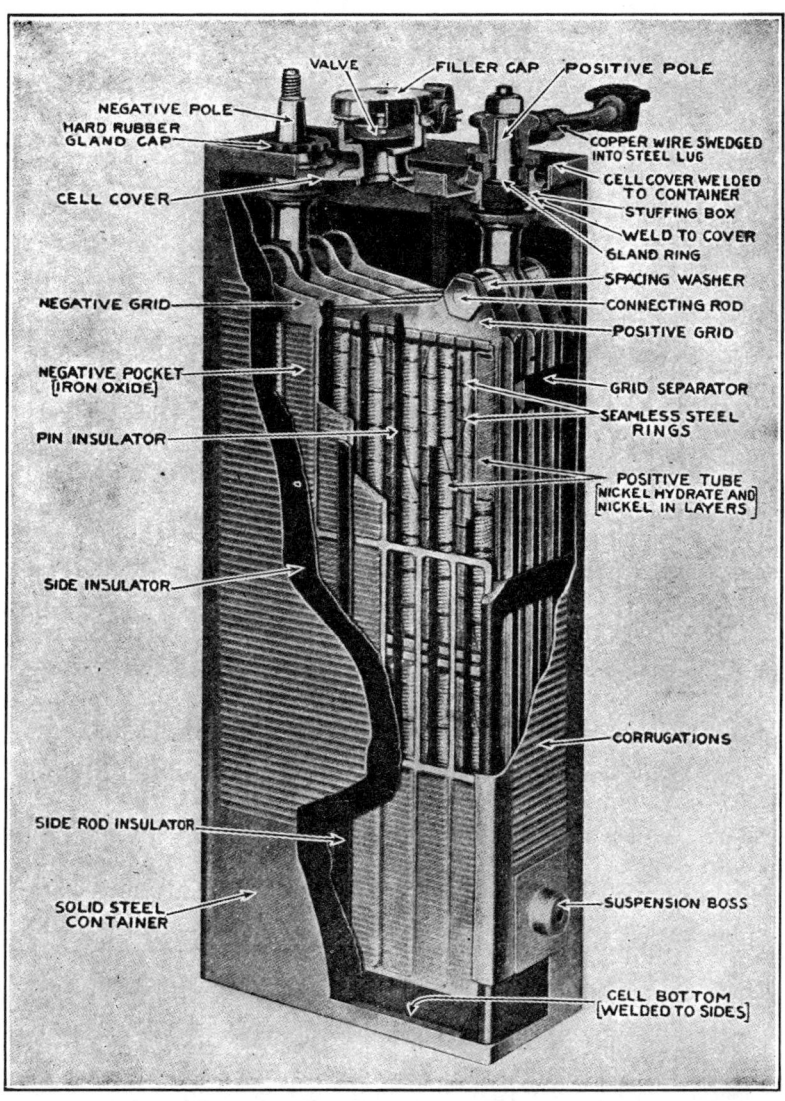

Fig. 18.—Part Sectional View of Edison Alkaline Battery, Showing Internal Arrangement of Plates and Relation of Principal Parts to Each Other.

cold-rolled steel, which is thoroughly welded at the seams and heavily nickel-plated, as shown at Fig. 18. The plates are assembled in positive and negative groups by means of threaded steel rods passing through holes in one corner of the plates and insulating washers. The terminal post is secured to the middle of the rod. The complete element or plate assembly stands on hard rubber bridges on the bottom of the can, and is kept out of contact with the sides of the container by hard rubber spacers attached to the end. The can cover is also of sheet steel, and contains fittings through which the electrodes pass, these being insulated from the cover by bushings of insulating material. A combined filling aperture and vent plug is secured to the center of the cover plate. The general arrangement of the Edison cell parts is clearly outlined at Fig. 18; the plate construction is depicted at Fig. 19, as well as the tubes from the positive plate and pockets used in the negatives.

Function of Separator.—Separators are necessary to keep plates of opposite polarity apart, and yet the space between the plates should be as small as possible in order to keep the internal resistance of the cells to a low point. It is apparent that current used to overcome internal resistance cannot be used in the external circuit. Separators may be of rubber, glass or wood. Perforated rubber sheets have been used, but these are not considered as good as wood separators, and are usually used in connection with them. Glass separators are used only in the largest cells, and usually consist of a series of vertical rods between the plates. Wood is used on all small cells. The material is specially selected and chemically treated. They are made very thin, and after cutting from seasoned wood they receive treatment to remove any elements that might cause damage if left in the wood. Thereafter, the separator strips are kept soaking in a weak electrolyte solution until they are installed in batteries. They must never be allowed to dry out, and even in transit from factory to service station they should be packed in such a way as to retain their moisture.

In impressing this matter on users of their batteries, the U. S. L. and H. Co. gives the following suggestions: "The owner must likewise do his part with the water cure as outlined above, to

FUNCTION OF SEPARATOR   51

prevent the separators from drying out in service. Once dried out a separator can never again, with or without water, be the same, but loses its vitality and is prone to split and undermine the battery's health. A badly shattered separator, of course, invites a direct short circuit, with resultant internal discharge of the cell. But battery plates seem eager to get together, and even a split in

Fig. 19.—Positive and Negative Plates Used in the Edison Alkaline Battery.

a separator affords the opportunity for 'treeing' across from negative to positive. That is, a foliage-like formation develops on the negative and extends through any available opening until it reaches the positive, and the short circuit thus produced not only dissipates the energy of the battery but to a greater or less degree cripples the battery. It also furthers the possibility of sulphation. The necessity for high capacity in starter batteries within little space demands that the plates shall be but a short distance apart. Thus, unless prevented by special provision, 'treeing' would occur across the bottoms of the separators. In assembling U. S. L. batteries pains are taken to make the separators of such length and to so fit them in place that their bottoms shall extend below the plates. Trouble has been experienced from distortion of the grids and the chiseling off of the separator bottoms by the sharp plate edges in their distorted condition. Buckled plates are saucer-shaped, so that the old-fashioned square-cornered plates did the most chiseling with their corners. To minimize the effects of buckling, even though it be the result of abuse, U. S. L. plates are round-cornered."

CHAPTER III

Storage Battery Defects—Loss of Battery Capacity—Sediment in Cells—Sulphation, Cause and Cure—Causes of Plate Deterioration—Cadmium Readings—Making Electrolyte—Features of Edison Cell—Tools and Supplies for Repairing—Taking Down Exide Batteries—Taking Down Gould Sealed Cells—Disassembling Willard Battery—Lead-Burning Apparatus—Lead-Burning Process—Battery Defects and Restoration Summarized.

Storage Battery Defects.—The subject of storage battery maintenance was thoroughly covered in a paper read by H. M. Beck before the S. A. E. and published in the transactions of the society. Some extracts from this are reproduced in connection with notes made by the writer and with excerpts from instruction books of battery manufacturers in order to enable the reader to secure a thorough grasp of this important subject without consulting a mass of literature. Endeavor has been made to simplify the technical points involved and to make the exposition as brief as possible without slighting any essential points. In view of the general adoption of motor starting and lighting systems on all modern automobiles, the repairman or motorist must pay more attention to the electrical apparatus than formerly needed when the simple magneto ignition system was the only electrical part of the automobile. The storage battery is one of the most important parts of the modern electrical systems, and all up-to-date repairmen and electricians must understand its maintenance and charging in order to care for cars of recent manufacture intelligently, as well as being able to understand the many industrial uses considered briefly in this volume.

In taking care of a storage battery, there are four points which are of the first importance:

First—The battery must be charged properly.
Second—The battery must not be overdischarged.

54   STORAGE BATTERIES SIMPLIFIED

Third—Short circuits between the plates, or from sediment under them, must be prevented.

Fourth—The plates must be kept covered with electrolyte, and only water of the proper purity used for replacing evaporation.

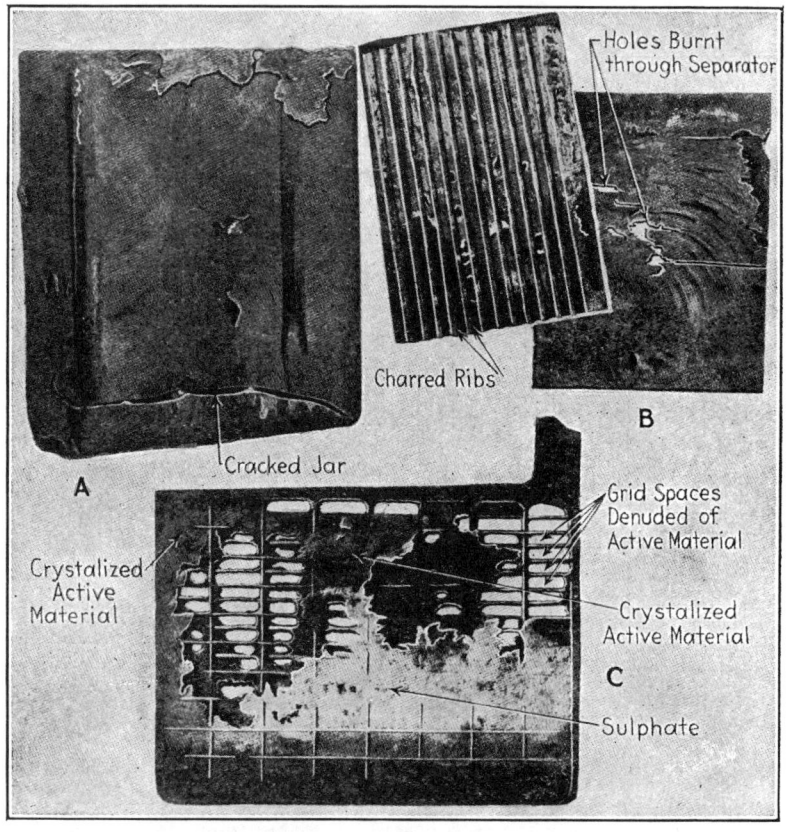

Fig. 20.—Defective Parts of Automobile Lighting Battery Abused in Service. A—Cracked Hard Rubber Cell Jar. B—Burnt Wood Separator. C—Badly Damaged Plate.

In the event of electrical trouble which may be ascribed to weak source of current, first test the battery, using a low-reading voltmeter. Small pocket voltmeters can be purchased for a few

STORAGE BATTERY DEFECTS 55

dollars and will be found a great convenience. Cells may be tested individually and as a battery. The proper time to take a reading of a storage battery is immediately upon stopping or while the engine is running. A more definite determination can be made than after the battery has been idle for a few hours and has recuperated more or less. A single cell should register more than two volts when fully charged, and the approximate energy of a three-cell battery should be about 6.5 volts.

If the voltage is below this the batteries should be recharged and the specific gravity of the electrolyte brought up to the required point. If the liquid is very low in the cell new electrolyte should be added. To make this fluid add about one part of chemically pure sulphuric acid to about four parts of distilled water, and add more water or acid to obtain the required specific gravity, which is determined by a hydrometer. According to some authorities the hydrometer test should show the specific gravity of the electrolyte as about 1.208 or 25 degrees Baumé when first prepared for introduction in the cell, and about 1.306 or 34 degrees Baumé when the cell is charged.

The following table gives the corresponding specific gravities and Baumé degrees:

Baumé	Specific Gravity	Baumé	Specific Gravity
0	1.000	18	1.141
1	1.006	19	1.150
2	1.014	20	1.160
3	1.021	21	1.169
4	1.028	22	1.178
5	1.035	23	1.118
6	1.043	24	1.198
7	1.050	25	1.208
8	1.058	26	1.218
9	1.066	27	1.228
10	1.074	28	1.239
11	1.082	29	1.250
12	1.090	30	1.260
13	1.098	31	1.271
14	1.106	32	1.283
15	1.115	33	1.294
16	1.124	34	1.306
17	1.132	35	1.318

The appended conversion formula and table of equivalents will be found of value in changing the reading of a hydrometer, or acidometer, from terms of specific gravity to the Baumé scale or vice versa.

$$\text{Sp. Gr.} = \frac{145}{145 - \text{Baumé degrees}} \text{ at } 60° \text{ F.}$$

Either voltage or gravity readings alone could be used, but as both have advantages in certain cases, and disadvantages in others, it is advisable to use each for the purpose for which it is best fitted, the one serving as a check on the other. Voltage has the great disadvantage in that it is dependent upon the rate of current flowing. Open-circuit readings are of no value, as a cell reads almost the same discharged as it does charged.

Loss of Battery Capacity.—When a battery gives indication of lessened capacity it should be taken apart and the trouble located. If the cell is full of electrolyte it may be of too low specific gravity. The plates may be sulphated, due to lack of proper charge or too long discharge. The cells may need cleaning, a condition indicated by short capacity and a tendency to overheat when charging. Sometimes a deposit of sediment on the bottom of the cell will short circuit the plates. If the specific gravity is low and the plates have a whitish appearance, there being little sediment in the cells, it is safe to assume that the plates are sulphated. Sediment should be removed from the cells and the plates rinsed in rain or distilled water to remove particles of dirt or other adhering matter.

Sediment in Cells.—The rate at which the sediment collects depends largely upon the way a battery is handled, and it is therefore necessary to determine this rate for each individual case. A cell should be cut out after, say, fifty charges, the depth of sediment measured and the rate so obtained used to determine the time when the battery will need cleaning. As there is apt to be some variation in the amount of sediment in different cells, and as the sediment is thrown down more rapidly during the latter part of a period than at the beginning, it is always advisable to allow at least one-fourth inch clearance. If the ribs in the bottom of the jars are 1¾ inches high, figure on cleaning when the sediment reaches a depth of 1½ inches. Before dismantling a battery

SEDIMENT IN CELLS

for "washing," if practical, have it fully charged. Otherwise, if the plates are badly sulphated, they are likely to throw down considerable sediment on the charge after the cleaning is completed.

There have been many complaints of lack of capacity from batteries after washing. Almost without exception this is found to be due to lack of a complete charge following the cleaning. The plates are frequently in a sulphated condition when dismantled, and in any case are exposed to the air during the cleaning process, and thus lose more or less of their charge. When reassembled, they consequently need a very complete charge, and in some cases the equivalent of the initial charge, and unless this charge is given the cells will not show capacity and will soon give trouble again. This charge should be as complete as that described elsewhere in connection with the initial charge.

Dangers of Flushing.—"Flushing," or replacing evaporation in cells with electrolyte instead of water, is a most common mistake. The plates of a storage battery must always be kept covered with electrolyte, but the evaporation must be replaced with pure water only. There seems to be a more or less general tendency to confuse the electrolyte of a storage battery with that of a primary cell. The latter becomes weakened as the cell discharges and eventually requires renewal. With the storage battery, however, this is not the case, at least to anything like the same degree, and unless acid is actually lost through slopping or a broken jar it should not be necessary to add anything but water to the cells between cleanings. Acid goes into the plates during discharge, but with proper charging it will all be driven out again, so that there will be practically no loss in the specific gravity readings, or at least one so slight that it does not require adjustment between cleanings. Thus, unless some of the electrolyte has actually been lost, if the specific gravity readings are low, it is an indication that something is wrong; but the trouble is not that the readings are low, but that something is causing them to be low, and the proper thing to do is to remove the cause and not try to cover it up by doctoring the indicator. The acid is in the cells, and if it does not show in the readings it must be in the form of sulphate, and the proper thing to do is to remove the cause of the sulphation if there is one, and then, with proper charging, drive the acid out of the plates and the specific gravity readings will then come back to the proper point. The too-frequent practice in such cases is to add electrolyte to the cells in order to bring up the readings which, as already explained, are only the indication of the trouble, and this further aggravates the condition, until finally the plates become so sulphated that lack of capacity causes a complaint. This practice of adding electrolyte to cells instead of water seems to be becoming more and more common.

Sulphation, Cause and Cure.—When plates are sulphated, to restore them to their original condition it is necessary that the battery be given a long, slow charge at about a quarter or a third of the normal charging rate. This should be continued until the electrolyte has reached the proper specific gravity and the voltage has attained its maximum.

It should be understood that sulphating is a normal as well as an abnormal process in the charge and discharge of storage batteries, and the difference is in the degree, not the process. The abnormal condition is that ordinarily referred to by the term. In normal service sulphating does not reach the point where it is difficult to reduce, but if carried too far, the condition becomes so complete that it is difficult to reduce and injury results. A very crude method of illustrating the different degrees of sulphating is to consider it as beginning in individual particles uniformly distributed throughout the active material. Each particle of sulphate is then entirely surrounded by active material. The sulphate itself is a non-conductor, but, being surrounded by active material, the current can reach it from all sides and it is easily reduced. This is normal sulphate. As the action goes further the particles of sulphate become larger and join together and their outside conducting surface is greatly reduced in comparison with their volume, so that it becomes increasingly difficult to reduce them, and we have abnormal sulphate.

Slow Charge Cures Sulphation.—The general cure for sulphating is charging, so that a cell, having been mechanically restored, the electrical restoration consists simply in the proper charging.

SLOW CHARGE CURES SULPHATION

Sulphate reduces slowly, and on this account it is a good plan to use a rather low current rate. High rates cause excessive gassing, heating, and do not hasten the process appreciably, so that it is the safer as well as the more efficient plan to go slowly. A good rate is about one-fifth normal. The length of charge will depend upon the degree of sulphating. In one actual case it required three months' charging night and day to complete the operation, but this was, of course, an exceptional one. The aim should be to continue until careful voltage and gravity readings show no further increase for at least ten hours and an absolute maximum has been reached. In serious cases it may be advisable to even exceed this time in order to make absolutely sure that all sulphate is reduced, and where there is any question it is much safer to charge too long rather than to risk cutting off too soon. A partial charge is only a temporary expedient; the cell, still being sulphated, will drop behind again.

The Gould Storage Battery Company give a number of characteristics by which cells in poor condition may be recognized in addition to the usual hydrometer readings. The common causes of plate deterioration are also given. The plates may be of poor color; the color of a wet positive plate in good condition varies from a rich dark brown (almost black) if the plate is fully charged to a reddish, fairly dark brown if discharged. A light grayish coating on the positive plates is not a bad indication, if by rubbing with a clean stick or piece of hard rubber a good color is evident immediately under the surface. The color is much lighter for dried plates. The wet negatives are of a light slate gray if charged and somewhat darker if discharged. When dry they are considerably lighter, and may even be somewhat yellowish if allowed to heat in drying. If the color of the plates is not as described they are probably considerably sulphated. If the cell voltage is markedly lower on discharge or higher on charge than it should be, sulphating is also indicated. If the acid strength is low, the cell should be investigated for short circuits or sulphated plates. Always be sure that the sediment does not touch the plates. It must be removed as soon as there is danger of this occurring.

Causes of Plate Deterioration.—Plates may get in poor condition from the following causes: 1st, Impure Electrolyte.—Either a poor quality used at the start, or through the use of impure water or through foreign substances getting into the cells. The remedy in this case, if the plates are physically in fair condition, is to replace the old electrolyte with new, the cells being in a discharged condition, and then thoroughly charge the battery. 2nd, Short Circuits.—These are not frequent if the sediment is removed before it touches the plates, as the wash of the electrolyte in most vehicle batteries resulting from the movement of the car would tend to free them. If they do occur, the cell should be completely dismantled, the plates straightened and the cell assembled again, the separators being completely replaced. The cell should then be thoroughly charged. 3rd, High Temperature.—At temperatures above 100° F. corrosion is quite rapid, and this limit must not be exceeded. If possible, the temperature should not exceed 90° F. The positives may be sulphated considerably from this cause and the plates grown abnormally and distorted. If they are thoroughly corroded they must be replaced; if not, they should be straightened and thoroughly charged. The conditions should be changed so that the battery will not again be subjected to the high temperature. 4th, Standing Discharged.—The positive plates especially may be badly sulphated from this cause. The indication of this condition is a light color of the positive plates, possibly with blotches of a grayish color. The remedy in this case is also complete charge, though care must be taken that too much active material is not thrown off during the charge. Under these conditions the active material is granular and non-cohesive, so care is needed in charging and discharging to restore the plates to efficiency.

Value of Cadmium Readings.—It is possible to make tests to determine the relative capacity of the positive and negative plates of any cell by means of a neutral electrode. Cadmium is well adapted for this purpose, and while such tests are not ordinarily made with small batteries they are very useful in determining the condition of large cells. The cadmium element consists of a stick of that material about the size of a lead pencil and 6 inches

VALUE OF CADMIUM READINGS

long, which is inserted into a soft rubber tube. Before inserting the cadmium stick, the rubber tube should be perforated with a number of holes about 1/16 of an inch in diameter. The rubber tube should extend at least $\frac{1}{8}$ of an inch beyond one end of the cadmium element. A small flexible rubber-covered copper wire should be soldered to one end of the cadmium stick. This wire may be joined to the negative pole of a voltmeter. A regu-

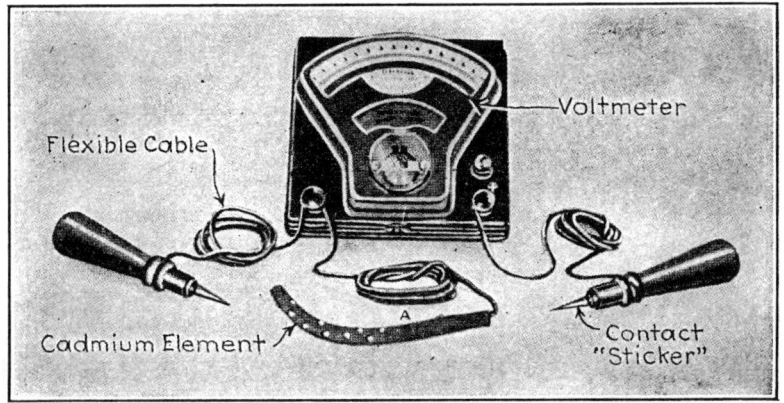

Fig. 21.—Apparatus Used in Making Cadmium Tests.

lar contact "sticker," or "stabber," such as shown at Fig. 21, is connected to the positive pole of the voltmeter.

After the total cell voltage throughout the battery is taken in the regular way, cadmium readings can be taken by inserting the rubber-covered end of the cadmium into the acid of the first cell, being sure that the bare cadmium stick does not touch either of the plate groups. Press the "sticker" leading to the positive pole of the voltmeter against the positive cross bar of the cell and note the reading. The difference between the reading of the positive group and the cadmium and the total cell voltage already taken will represent approximately the negative cadmium reading. For example, if the positive cadmium reading is two volts and the regular cell voltage 1.85, the negative cadmium would be 0.15. A positive is discharged when its cadmium reading at its regular catalog discharge rate is about 1.95, and the negative when the cadmium reading is approximately 0.25 volt. At these values the cell voltage would be 1.7.

As a cell discharges the positive cadmium reading decreases, while the negative cadmium reading will increase. The curve shown in Fig. 22 illustrates the discharge of a cell in which the positive plates are low in capacity. Note the rapid drop in cell and positive cadmium voltage after three hours but the slow rise

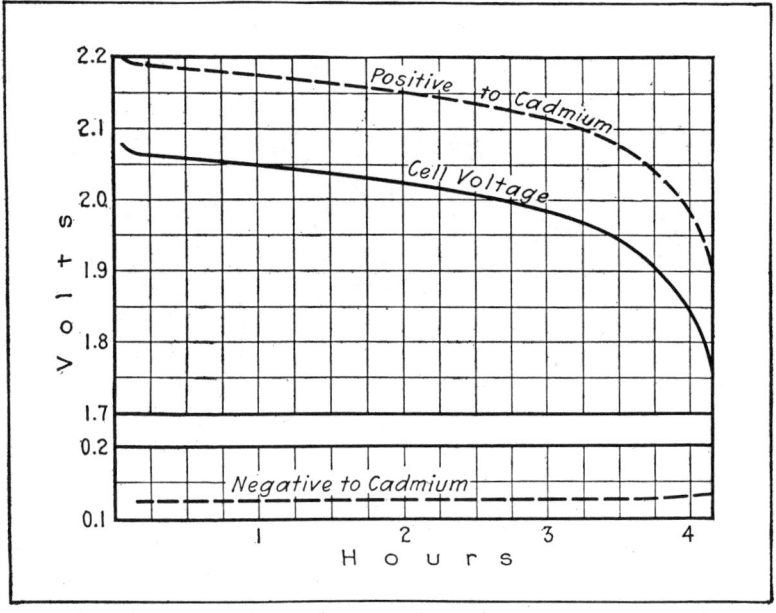

Fig. 22.—Curves Illustrating the Discharge of a Cell During a Cadmium Test in Which the Positive Plates are Low in Capacity.

in the negative cadmium voltage. All voltage readings must be taken while the battery is discharging at its catalog discharge rate. The readings should be carefully filed and compared with previous tests. In this way any cell not in condition will be readily found.

Making Electrolyte.—Electrolyte, as used in all lead plate types of batteries, consists of a mixture of pure sulphuric acid and

MAKING ELECTROLYTE

distilled or other pure waters. Concentrated sulphuric acid is a heavy, oily liquid, having a specific gravity of about 1.835. A battery will not operate if the acid is too strong, and it is therefore diluted with sufficient pure water to bring it to a gravity of 1.270 to 1.300 for a fully charged battery. Stronger electrolyte than this is injurious. To prepare electrolyte from sulphuric acid of 1.835 specific gravity, mix with water in the proportions indicated in Fig. 23 for the desired specific gravity, taking the following precautions:

Use a glass or earthenware vessel, never metallic.

Carefully pour the acid into the water, never water into acid.

Stir thoroughly with wooden paddle and allow to cool before reading the gravity.

Both the water and the sulphuric acid used in making electrolyte should be chemically pure to a certain standard. This is the same standard of purity as is usually sold in drug stores as "CP" (chemically pure), or by the chemical manufacturers as "battery acid."

Electrolyte made from sulphuric acid meeting the following specifications will be satisfactory:

Sulphuric acid to be high grade, either the so-called "Brimstone Oil of Vitriol" or "Contact Process Acid" made from sulphur of good quality. Must be water white in color and show no sediment on standing.

By analysis, impurities must not exceed the following:

Platinum	None
Arsenic	Trace
Manganese	Trace
Iron	0.005%
Chloride	0.001%
Nitrogen in any form	0.01 %
Copper	0.002%
Sulphurous acid	None
Organic matter	None

Must be free from all substances other than stated above.

As great care is exercised in the manufacture of storage battery plates and in the furnishing of acid to secure a high degree of purity, obviously attention should be paid to the purity of the water used both in the dilution of concentrated acid, if this is

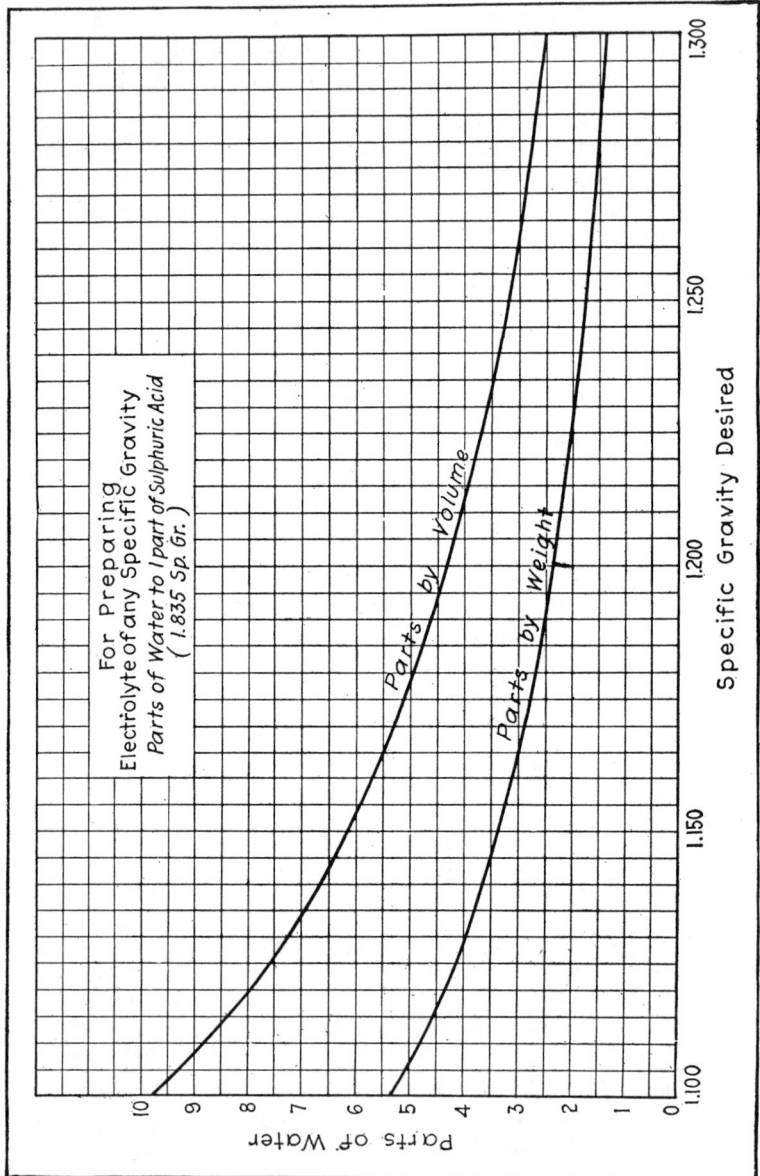

Fig. 23.—Chart for Preparing Electrolyte of Any Specific Gravity.

MAKING ELECTROLYTE

attempted, and replacing the loss in cells occasioned by evaporation and overcharge. The frequent addition of water required to replace evaporation leads eventually to a concentration of any impurities which it may contain. As ordinary water supplies are not pure, their use is always questionable. Water from natural sources should be used only with the approval of competent chemists. Rain water should not be used if distilled water is available, as it often contains traces of nitric acid and ammonia, either of which is harmful to storage batteries. Distilled water is preferable and should always be used unless otherwise advised by the chemists. The water should be stored in carboys or thoroughly cleaned whiskey barrels. Water obtained by condensing the exhaust from engines and which may thus contain cylinder oil and other impurities should never be used for battery purposes. In cleaning batteries the ordinary tap water may be used provided it does not contain a great quantity of impurities.

In this connection, the expression "chemically pure" acid is often confused with acid of "full strength." Acid may be of full strength (approximately 1.835 sp. gr.) and at the same time chemically pure. If this chemically pure acid of full strength be mixed with chemically pure water, the mixture would still be chemically pure, but not of full strength. On the other hand, if a small quantity of some impurity be introduced into chemically pure acid, it would not materially reduce the strength, but would make it impure. The usual method of determining the strength of electrolyte is by taking its specific gravity. The method is possible on account of the fact that sulphuric acid is heavier than water. Therefore the greater the proportion of acid contained in the electrolyte the heavier the solution or the higher its specific gravity. By specific gravity is meant the relative weight of any substance compared with water as a basis. Pure water, therefore, is considered to have a specific gravity of 1, usually written 1.000 and spoken of as "ten hundred." One pound of water is approximately one pint. An equal volume of concentrated sulphuric acid (oil of vitriol) weighs 1.835 pounds. It therefore has a specific gravity of 1.835 and is spoken of as "eighteen thirty-five."

Since electrolyte, like most substances, expands when heated, its specific gravity is affected by a change in temperature. For the convenience of the operator, the following table is given to show the variation in the electrolyte specific gravity at various temperatures likely to be met with in service:

Degree Fahrenheit	Specific Gravity
106	1.208
97	1.211
88	1.214
79	1.217
70 normal	1.220 normal
61	1.223
52	1.226
43	1.229
34	1.232
25	1.235
16	1.238
7	1.241

The solution must be allowed to stand several hours to cool. Never add hot or even warm electrolyte to a cell, as the plates are liable to be dangerously sulphated thereby. The strength of the resultant solution should always be checked by hydrometer readings reducing the latter to 70 degrees Fahr.

Features of the Edison Cell.—The instructions given apply only to batteries of the lead plate type and not to the Edison battery, which is entirely different in construction. The Edison cell uses an electrolyte consisting of 21% solution of potash in distilled water so that the electrolyte is alkaline instead of acidulous. For 6-volt ignition and lighting service it is necessary to use 5 cells owing to the lesser voltage of the Edison batteries. The average voltage during discharge is but 1.2 volts per cell, and is not as constant as is the case with a lead battery, the voltage of which may be as high as 2.5 volts per cell.

An Edison 6.5-volt battery used for lighting or ignition may be charged completely in ten hours. A feature of the Edison battery is that overcharging at the normal rate has no harmful effects, and it is advised by the maker to give the battery a 12-hour charge once every 60 days or when the electrolyte is replenished. The electrolyte must be kept sufficiently high so as to

FEATURES OF THE EDISON CELL 67

cover the plates, and any loss by evaporation must be compensated for by the addition of distilled water. Another feature in which the Edison battery is superior to the lead plate type is that the plates will not be injured if the cells are allowed to stand

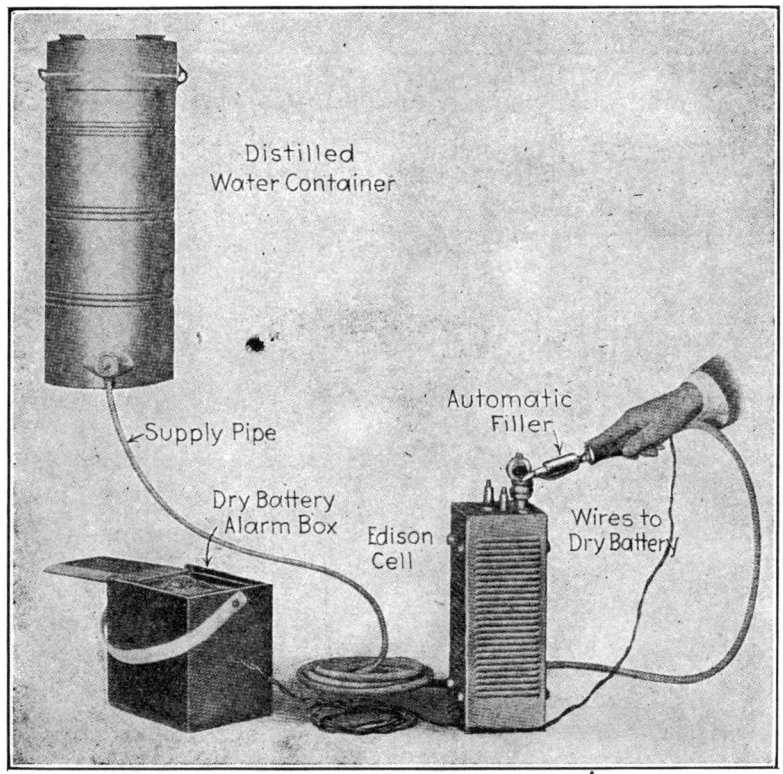

Fig. 24.—Automatic Filler for Renewing Water Supply in Edison Alkaline Cell.

in a discharged condition. The external portions of the cells must be kept clean and dry, because the container or can is made of a conducting material. The vent caps must be kept closed except when replacing electrolyte or bringing the level up to the proper height by adding distilled water. Care should be taken to avoid short circuiting of the battery by tools or metal objects, and special emphasis is laid on the precaution that no acid or electrolyte containing acid be poured into the cells. It is said that the Edison battery has a longer life than the lead plate type of equal capacity.

Repairing Exide Sealed-Type Batteries.—The smaller Exide

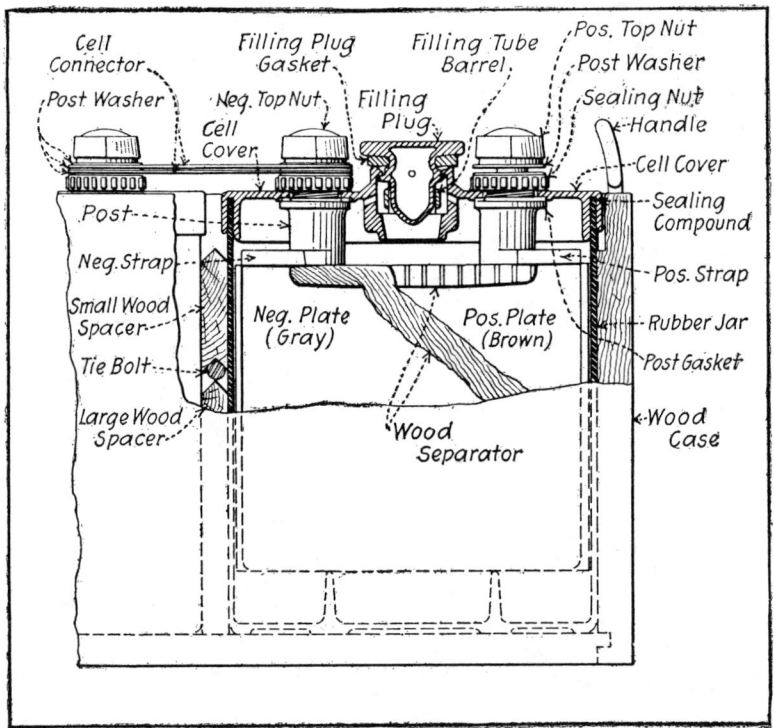

Fig. 25.—Exide Double Seal Bolted Connector Type Battery.

cells, such as used in electric vehicles and for automobile starting, lighting and ignition systems, are made in two types. In one a double flange cover is depended on to keep the electrolyte from splashing out, this construction being shown in the sectional view at Fig. 25. In the other a single-flange cover, as outlined at Fig. 26, is utilized in connection with sealing compound. The

REPAIRING EXIDE BATTERIES 69

double-flange cover has two downwardly projecting flanges, one fitting inside and the other outside of the cell jar. The two flanges form a channel or slot, holding the jar walls. In order to insure a tight joint a small amount of sealing compound is placed at the bottom of the slot and between the cell cover and the top of the rubber jar.

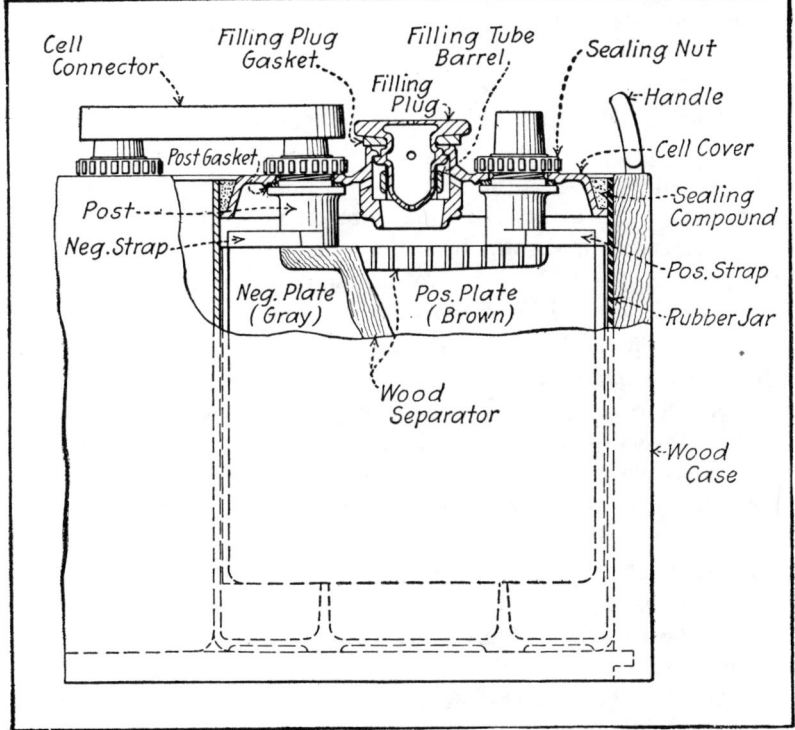

Fig. 26.—Exide Single Seal Burned Connector Type Battery.

To unseal this type of battery two stout boards about one-quarter of an inch longer than the height of the jar are needed. Rest the side flanges of the jar on these, as shown at Fig. 27 A, so the cell will be raised and the weight supported by the wood blocks. Next warm the cover around the edges so that the sealing compound will soften. Of course, the terminal straps and sealing nuts have been removed. As soon as the compound has softened, a little pressure with the thumbs on the terminal posts will cause the jar to drop out of the cover. The warming may be accomplished by passing a moderate flame quickly around the cover, taking care not to hold it at any one spot long enough to burn the rubber.

With the single-flange type the sloping sides of the cover and flange at the bottom give a space about a quarter of an inch wide for the sealing compound. To unseal this type of seal a flat-bladed putty knife or a screw-driver is heated in the flame and run through the sealing compound close to the jar wall all the way around. This will loosen the compound and the element can be lifted out of the jar. When taking the elements apart, i. e., separating the positive and negative plate groups, never allow them to stand in the air, but always put them in some weak electrolyte. If the cell jar is broken and the parts are in otherwise good condition it is not necessary to remove the sealing nuts on the single-flange type, as the entire assembly may be lifted back into a new jar. If the cover is cracked it will be necessary to loosen the sealing nuts and to supply a new cover.

If the plates are found to be buckled, the operation of straightening is relatively simple. Spacing boards of suitable thickness are placed between the plates, also outside of the plate group, and the whole is put in a vise, as shown at Fig. 27 B, and subjected to a gradual pressure. If, in addition to the buckling, the negatives have shed active material due to starvation or other abuse, it will be necessary to use a new set of plates. When the active material is very hard and not spongy it is "sulphated," and particular care should be taken in charging, after the cell is reassembled, to make sure that the electrolyte is brought to its maximum gravity.

It is well to examine the wood separators to see that the ribs are not worn off and that there are no splits or other perforations to reduce the mechanical strength or destroy its utility as a separator. The method of removing and inserting separators is clearly outlined at Fig. 27 C. It is always necessary to clean out cell jars thoroughly after the elements have been removed in order to clear out all sediment and fallen active material. In inserting separators it is well to note that the flat side of the wood goes against the negative plate and the rib side against the positive plate. Where perforated rubber sheets are used in addition to

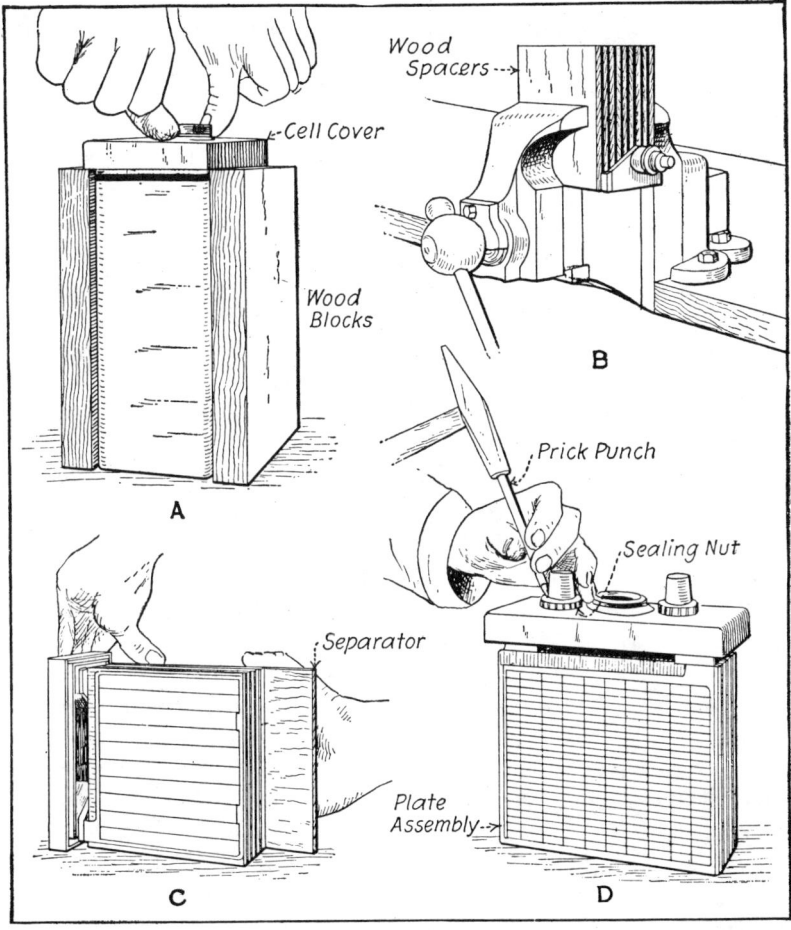

Fig. 27.—Some Operations When Repairing Exide Battery. A—Removing Double Seal Cover. B—Straightening Buckled Plates. C—How to Remove Wood Separators. D—Method of Locking Sealing Nuts on Burned Connector Type Cells.

the wood separators, these are always placed against the grooved side, one to each wood separator, and the two inserted between the plates together. When the separators are all in place the edges should be tapped with a wood block until they project equally on each side of the plates.

The sealing nuts are tightened with a special wrench, which is shown in Fig. 28. To lock the sealing nuts in place a prick-punch is used to burr the thread in spots above each nut, as shown at D, Fig. 27. This will slightly upset the alloy thread on the post and prevent the nut from coming loose. This is necessary only on those types of cells having burned connections, as in bolted connections the thread of the nut post does not extend above the sealing nut. The top, or clamping nut, acts as a lock to prevent the sealing nut from loosening.

Before sealing a cell, always wipe the surfaces against which the sealing compound is to be placed with ammonia and allow it to dry thoroughly. Otherwise the compound will not stick. In the double-flange type a string of sealing compound about 3/16ths inch in diameter is made by rolling some special compound between two boards. This is packed in the space between the flanges and is heated before the cell cover is pushed in place.

While it is not difficult to release those types of cells having bolted-on connecting strips, the burned-on type connectors cannot be removed as easily. To remove these solid lead links it is necessary to bore the connectors centrally over each post with a $5/8$-inch wood bit. Another method is to play a burning flame on the joint to soften the lead and then to pull off the connector with a pair of pliers. The method of taking down the bolted connection is clearly outlined at Fig. 28. The first step is to remove the filling plugs to provide more room for working on the battery terminals. A special socket wrench is provided for the alloy-covered top nuts, this being easily turned by a special ratchet wrench. After the top nut is removed the spacing washers and connector strips may be taken off, and it is well to take off the connector strips without bending them. It is also well to save the alloy washers. These are placed one above, one between and one below the two connector straps. Two types of connector strips

REPAIRING EXIDE BATTERIES

are provided, a simple form consisting of a straight piece, and a later type, in which the lead-covered copper connectors are provided with cast lead ends that eliminate the spacing washers.

Do not try to unscrew the sealing nuts with a Stilson wrench. The wrench teeth will not only damage the sealing-nut corrugations, but the pressure may squeeze a nut so tightly into the threads that it can be removed only with difficulty. The special

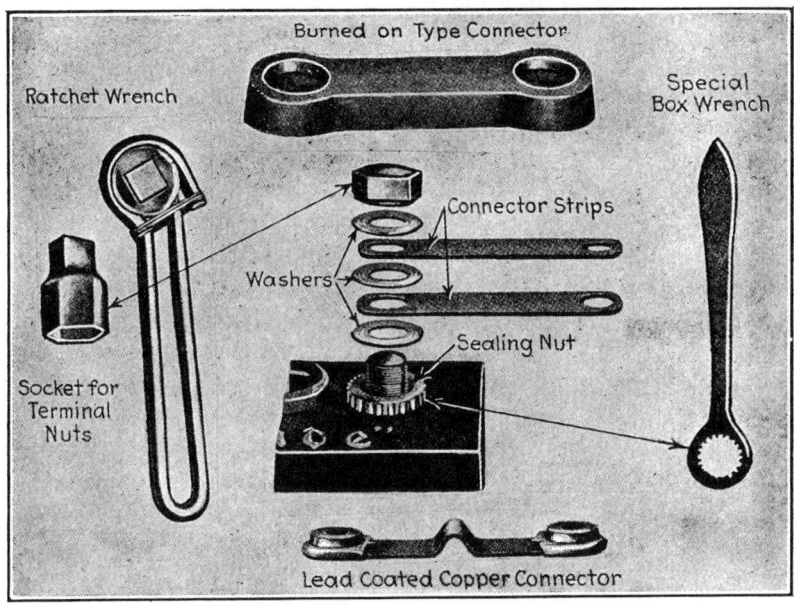

Fig. 28.—Tools Used and Method of Disassembling Bolted Type Exide Connectors.

box wrench illustrated is necessary for removing the sealing nuts without damage. This has a series of small projections which fit into the corresponding spaces on the sealing nut and make it very easy to turn that member. Whenever a bolted connection-type cell is assembled, the first step is to grease the studs well with vaseline. Slip one of the connector links over the posts of adjoining cells, then an alloy washer over each post, followed by a second connector and a second washer. The last washer is then put in place and the top nut pulled up tight with a properly fitting socket wrench. If a monkey wrench or worn S-wrench is used the soft corners of the alloy-covered nut are apt to be marred if the wrench slips.

Battery Repair Tools.—The following list of tools and apparatus is given in the Gould Instruction Book, and will be found of value in repairing storage batteries of any make: One pair of rubber gloves, to protect the hands from acid; one 7-inch end-cutting nippers or one pair bolt cutters, for cutting connectors, plate lugs, etc.; two pairs of combination pliers, for pulling elements from jars; one triangular lead scraper, for cleaning plate lugs, terminals, etc.; one putty knife, for removing sealing compound; one half-inch wide wood chisel, for the same purpose; one five-inch screw-driver, for removing sealing compound and covers; one single-end wrench for removing terminal nuts; several coarse files and handles, for filing plate lugs, straps, etc.; one steel wire brush for cleaning files and battery terminals; one ball pein hammer, medium size, for general work; one 10-inch ratchet bit brace, for drilling links loose from pillar posts; one $5/8$-inch diameter bit-stock drill, for removing $5/8$-inch connectors; one $7/8$-inch bit-stock drill, for removing $7/8$-inch connectors; one small drill, to start holes; one center punch, for centering terminals to drill; one adjustable hacksaw frame and three 8-inch blades to fit it; one iron ladle, for pouring sealing compound; one pair blue glasses, for use when lead burning; one soft-rubber bulb syringe, for flushing and equalizing electrolyte; one burning-rack, with extra guide plates; one hydrometer, for mixing electrolyte; one thermometer, for reading cell and electrolyte temperature, and one lead-burning outfit. A group of the tools recommended by the Electric Storage Battery Company for work on the Exide batteries is shown at Fig. 29. The special terminal and box wrenches shown will fit the Exide battery terminals only.

To Repair Gould Batteries of the Sealed Types.—Batteries to be repaired may have been in service but a short time, the necessity for repairs being broken terminals, leaky jar, plates in one cell short circuited, etc. Under these circumstances it is only

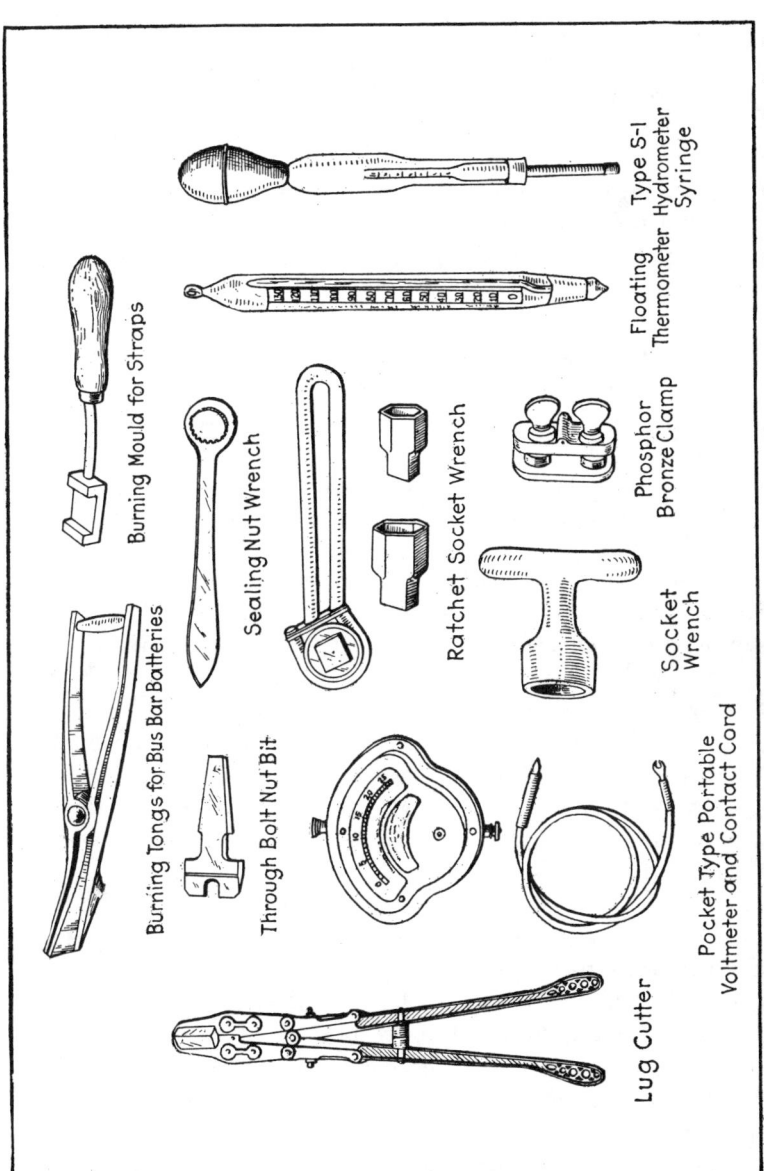

Fig. 29.—Tools and Apparatus Recommended by the Electric Storage Battery Company to Assist in Maintenance and Repair of Exide Batteries.

necessary to make the repairs indicated by test. If, however, the battery has been in service for six months or longer, and the cell or cells to be repaired show general deterioration, it is very probable that the other cells in the battery are in nearly as bad condition, and the elements should be removed from all of the jars in the battery. If, upon inspection, plates are still in good condition, the wood separators should be discarded and new wood separators put in the good cells as well as those requiring repairs.

To Dismantle a Cell: Have battery fully charged before dismantling a cell.

Remove vent cap and washer.

To remove terminal or connecting link, center the tops of terminals and connectors over the terminal posts with a center-punch and drill down to depth of 3/4 inch, using a 5/8-inch drill if you have 3/4-inch posts and a 7/8-inch drill if you have 1-inch posts. Terminals or links can then be removed by working back and forth gently with gas pliers.

To remove top cover. Soften the sealing compound by a jet of steam or a gas flame. The use of the flame requires very careful manipulation and continual attention of the operator. Care must be taken that the flame does not burn or scorch the edges of the cover. Then gently pry the cover from the jar.

With a heated putty knife or screw-driver, clean the compound from the inside edges of the rubber jar. The element can now be removed (with the lower cover) by grasping each terminal post firmly with gas pliers and pulling up slowly but strongly, holding the battery down meanwhile.

If separators are in good condition and a jar replacement only is to be made, set the element, with bottom cover, in electrolyte or water till ready to replace.

Remove the bottom cover from the element after cleaning compound away from the posts. The covers may have warped from the heat. If so they should be heated again by being placed in boiling water, straightened out and laid on a flat surface to cool.

Separate the positive and negative groups and discard the wooden separators. If rubber sheets also are used, those that are not broken should be washed and laid aside for future use. The

REPAIRING GOULD TYPE BATTERIES

negative group should be washed and also laid aside until needed. The negative group should be immersed in water, otherwise it takes up oxygen from the air, which is liable to cause dangerous heating.

To remove a leaky jar, pour boiling water in the jar to soften the surrounding compound and lift the jar from the case. If the compound cannot be softened with boiling water, use a jet of steam or a flame on the inside of the jar. To install a new jar, pour boiling water in the jar. When it is thoroughly heated, press it carefully into place.

To Replace an Element: To assemble the new element: inter-

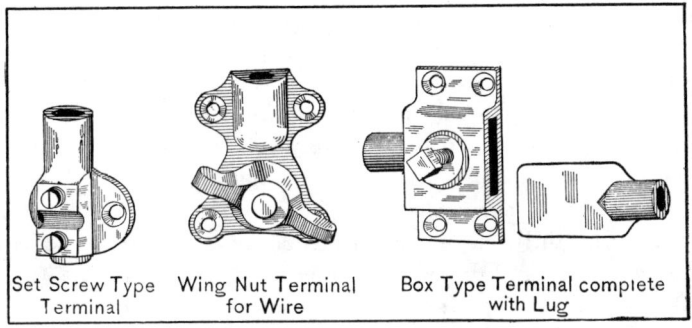

Fig. 30.—Terminal Types Used in Connection With Exide Vehicle Batteries.

mesh the positive and negative group, positive and negative plates alternating. As a negative group contains one more plate than does the positive, both outside plates will be negative. Lay the element on its side, and put the separator retainers in position. Insert the separators between each pair of plates. If wooden separators only are used, the grooved side of the separator should be next the positive plate. If wood separators and rubber sheets are used, they should be inserted together, the rubber sheet between the positive plate and the grooved side of the wood separator. See that the separators are against the retainers and that they extend equally on either side of the element.

Grasping the element by the pillar posts, lower gently into the jar. Fill with electrolyte of the proper density (see "Electrolyte") and let the cell cool for at least twelve hours. Develop the plates. It is advisable to develop with the cover off on account of better ventilation and greater convenience in taking thermometer and hydrometer readings. Furthermore, if a fault develops it can be remedied without having to remove the cover. Proceed as follows: Burn a copper wire (about No. 10) to the top of each terminal post with a few drops of burning material, just enough to make good connection. Connect these wires to the charging source. Develop at a rate equal to six-tenths (.6), the final rate of the battery. The time required to develop at this rate will be about sixty hours. After the developing has gone on for thirty hours, disconnect the charging wires and reconnect so as to charge the balance of the cells in the battery as well as the cell or cells being developed. When the cell voltage and the specific gravity have remained unchanged for five hours, the cell is fully developed. Even up the electrolyte in the cells to 1.300.

Place the upper and lower covers in boiling water. When the lower cover has become thoroughly heated, press it gently into position. Carefully clean the inside edges of the jar and cover with warm water and dry with a flame. Otherwise, the compound will not stick. If there are any openings between cover and jar wide enough to allow the heated compound to run into the jar, work a small amount of compound around these edges, using a putty knife or brush before starting to pour. Pour melted sealing compound around the edge. Allow a few moments for this compound to harden slightly, then pour melted compound until it reaches a point slightly above the top of the expansion chamber, and at once press the heated top cover on the compound. Place a weight on the top cover and let it remain until the top cover cools fast to the compound. Pour melted compound around the edges and to the level of the top of the cover and smooth off with heated putty knife. Burn the connecting links and terminals to the pillar posts.

Repairing Willard Automobile Type Batteries.—In repairing a Willard storage battery a definite routine must be followed in tearing down and building up same in order that it will be in

REPAIRING WILLARD TYPE BATTERIES

the best condition when reassembled. These steps are as follows:

First: Remove all vent plugs and washers.

Second: Center punch both top connectors in each cell which is to be repaired; then drill ¾ inch into top connector with a

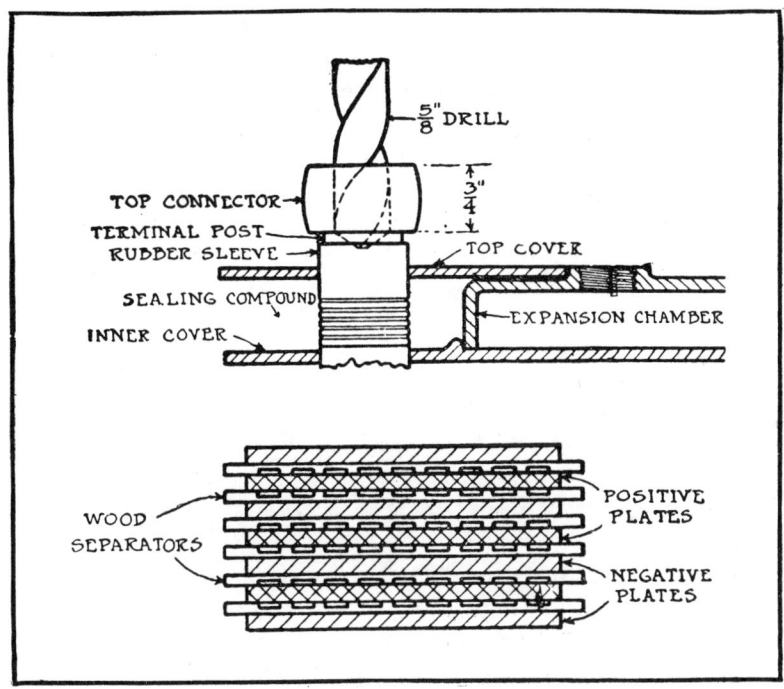

Fig. 31.—Diagram Showing Construction of Points to be Reached in Rebuilding or Tearing Down Willard Storage Battery.

⅝-inch diameter drill. Now pull off top connector with pair of pliers.

Third: Apply gas flame or blow-torch flame to the top of the battery long enough to soften the sealing compound under the top cover. Now, with heated putty knife, plow out the sealing compound around the edge of the top cover.

Fourth: Insert a putty knife, or any other thin, broad-pointed tool, heated in flame, along underside of top cover, separating it from the sealing compound. Then, with putty knife, pry the top cover up the sides and off of the terminal posts.

Fifth: Then, with heated putty knife, remove all sealing compound from inner cover.

Sixth: Now play the flame onto the inner cover until it becomes soft and pliable; then take hold of both terminal posts of one cell and remove the elements from the jar slowly; then lift the inner cover from the terminal posts.

Seventh: Now separate positive and negative elements, by pulling them apart sideways. Destroy old separators.

Eighth: To remove a leaky jar, first empty the electrolyte from the jar, and then play the flame on the inside of the jar until the compound surrounding it is soft and plastic; then, with the aid of two pairs of pliers, remove it from the crate, slowly, lifting evenly.

Ninth: To put in a new jar, in place of the leaky one, heat it thoroughly in a pail of hot water and force it gently.

Tenth: In reassembling the battery, first assemble the positive and negative elements, pushing them together sideways, then turn them on the side, and with both hold-downs in place, insert new separators, being very careful to have the grooved side of the separators next to each side of the positive plate. Also be careful to have the separators extend beyond the plates on each side, so there will be no chance of the plates short circuiting. Now press all separartors up against hold-downs.

Eleventh: Heat up inner cover with the flame; then place same on terminal posts; then take hold of both terminal posts and slowly lower the elements into the jar.

Twelfth: Now, with expansion chamber in place on the inner cover, until it reaches the level of the hole in the top of the expansion chamber, i. e., so that when the top cover is replaced it will squeeze the sealing compound off the top of the expansion chambers.

Thirteenth: Now soften top cover with flame and replace on terminal posts until it rests on top of expansion chamber; then place a weight on top cover until sealing compound cools.

Fourteenth: Now pour sealing compound around the edge of

REPAIRING WILLARD TYPE BATTERIES

the top cover until it reaches the top of top cover; then, when the sealing compound has cooled, take a putty knife and scoop extra sealing compound off of top cover, making a smooth surface over all the top of the battery.

Fifteenth: In burning the top connector to terminal post, proceed as follows: Scrape the hole of the top connector until the surface is bright and clean; scrape terminal post until top and edge of all surfaces are free of dirt. Now, scrape a piece of lead thoroughly, preferably a small bar; then apply hydrogen-gas flame, mixed with air under pressure, to the top connector and terminal post assembled, at the same time heating lead bar. When top connector and terminal post begin to melt, apply lead bar directly on same, melting it, thus making a firm burned connection. Then fill rest of hole-space with melted lead and smooth off even with top of top connector.

General Care of All Lead Batteries: The battery boxes must be kept clean and dry. The acid-proof paint of both the boxes and the tanks must be kept in good condition by repainting when necessary. The terminals must be kept thoroughly clean and covered by a coating of vaseline. Corroded copper, iron or any other foreign materials must not be allowed to get into the cells. If, through accident, this occurs, the acid in such cells must be thrown away and new electrolyte used. Matches or exposed flames of any kind must not come near the battery boxes, especially when the cells are charging. The gases thus given off are explosive when sufficiently concentrated. Temperatures higher than 100° F. are to be avoided, as the corrosion of the positive plates is accelerated. Low temperatures are not injurious, although they temporarily reduce the capacity of the battery.

Lead-Burning Outfits.—In all batteries having permanently jointed connections the various joints are produced by melting of a portion of the parts to be joined by a process termed "lead-burning," and forming a solid weld by means of heat to melt the lead, which may be produced with illuminating gas, hydrogen gas or the electric arc. The illuminating gas outfit is the simplest and can be used to advantage wherever that gas is available. It consists of a special-burning tip and a mixing valve. A supply of compressed air is necessary, the pressure ranging from 5 to 10 pounds, and the various parts of the apparatus are joined together by 5/16-inch rubber hose, which is securely wired to the apparatus to insure tight connection. This is made necessary by the air pressure advised. The mixing valve is a very simple fitting, comprising of two shut-off cocks attached to a common outlet pipe. One of the cocks regulates the gas supply, the other controls the amount of air. Naturally, the mixed air and gas issue from the outlet pipe. The burner is a special form, which gives a very hot flame. When the flame is properly adjusted for burning it will have a greenish color. If there is too much gas, the flame will be yellow and be very ragged. If the flame is a blue color, gradually becoming less visible, too much air is provided, and as

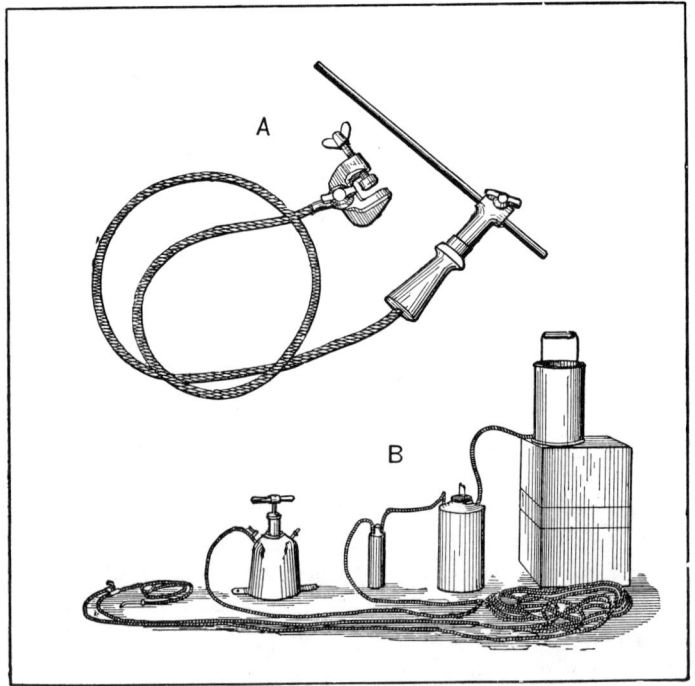

Fig. 32.—Lead-Burning Outfits. A—Electric Arc Set. B—Hydrogen Gas Generator.

LEAD-BURNING OUTFITS

a result it is lacking in heating power. The hottest part of a properly adjusted flame is just past the end of the inner point. Do not hold the flame too near the work, as the heating effect of the flame will be diminished if it is spread. If the air pressure is obtained from a tank holding a supply for blowing up automobile tires, for instance, a reducing valve must be introduced in the air line between the tank and the burner. The best method of producing the air pressure is by a small blower.

The apparatus needed for arc burning is shown at Fig. 32 A. The advantage of this method is that current from a six-volt battery may be used, not requiring the fitting of auxiliary apparatus. Although called an "arc-burning outfit," it is said that the best results are obtained by using the carbon as a soldering iron after it becomes heated without actually drawing an arc. The outfit is very simple, consisting of a carbon holder with cable, a clamp and a number of ¼-inch diameter carbons. The method of using it for reburning connectors is easily understood. The connector to be burned is connected to one terminal of the storage battery by a piece of cable, which can be made fast to the latter by means of a clamp. It is essential that the contact surfaces be scraped bright to secure a good electrical connection. The cable attached to the carbon holder is connected to the other battery terminal. If a battery is partially discharged the three cells will be needed, but if the battery is fully charged three cells may give too much voltage. The amount of current passing through should be sufficient to raise the temperature of the carbon to at least a bright cherry red while it is in contact with the joint. The carbon should be sharpened to a long point and should not project from the holder more than two or three inches. The holder should be cooled off occasionally by plunging it into a pail of water. After being used for a short time the carbon will not heat properly because of a scale film produced on the surface. This should be cleaned off till the bare carbon is exposed before proceeding with the work.

The hydrogen-gas outfit, such as shown at B, Fig. 32, while more expensive and troublesome than the illuminating gas burner, produces a much superior flame for lead burning, and is very generally used where a large amount of work is done. The hydrogen outfit shown is supplied by the Electric Storage Battery Company, and consists of the following parts: One generator; one washbottle; one air pump and tank combined; one branch pipe; one finger pipe and set of tips; one 50-foot length 5/16-inch rubber tubing; one two-foot length 3/4-inch rubber tubing; two rubber stoppers; one triangular scraper. The material for charging is: zinc, 15 pounds; water, 12 quarts; sulphuric acid, 2½ quarts.

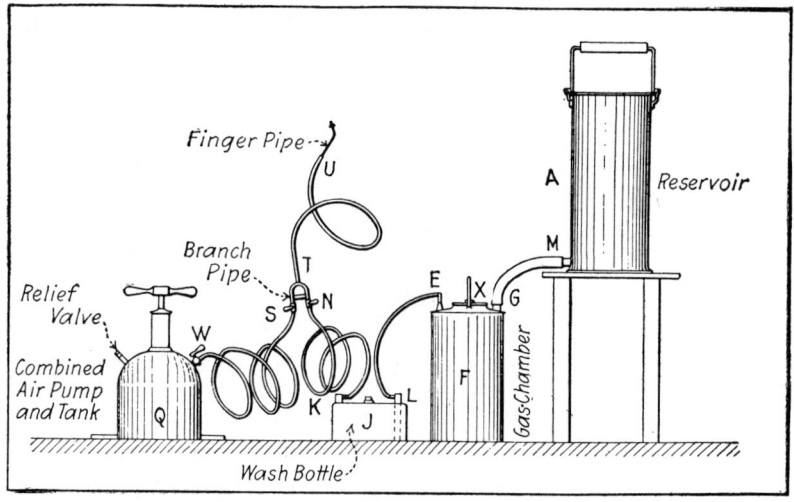

Fig. 33.—How Hydrogen Gas Generating Outfit is Set Up.

The apparatus is connected up as shown at Fig. 33. The instructions for using are sent with each outfit, but a brief outline of the method of joining the parts may be of some value. The bottom of the reservoir A must be higher than the top of the gas chamber F. Connect the lower outlet M of the reservoir A with the pipe G, coming out of the top of the gas chamber F. Put a short piece of 5/16-inch hose on the outlet E coming from the gas chamber F, and kink this hose to constrict the passage and prevent anything coming through it. Put a rubber stopper in outlet H of gas chamber F, and inspect carefully to see that it is tightly in place. Remove the hand-hole cover X from the top of the gas chamber, place a quantity of zinc on the grating. Next replace the hand-hole cover, making sure that it is securely fastened, and screw down tight on its gasket. An amount of water is placed in reservoir A and then a certain amount of vitriol is poured into the water. The wash-bottle J is filled half full of water, and its outlet K is connected to one side N of the branch pipe. The other side of the branch pipe S is joined to the outlet W on the air tank Q. The finger pipe U is connected with

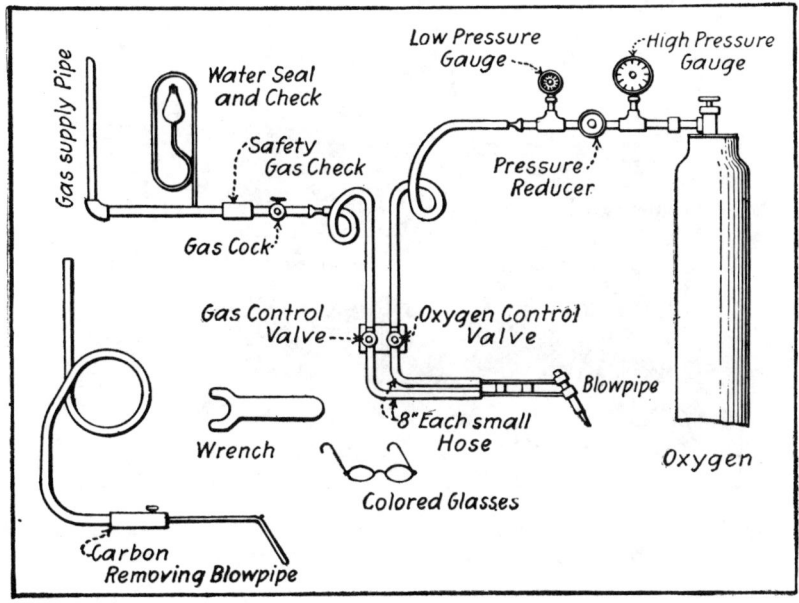

Fig. 34.—Method of Using Oxygen-Illuminating Gas Blow-Pipe Outfit.

the outlet T of the branch pipe. Both cocks S and M are closed. Next take the kink out of the hose connected to outlet E of the gas chamber and allow the air to escape until the charge of water and vitriol runs down from the reservoir into the gas chamber, then slip the free end of this hose over the outlet marked L on the wash-bottle.

As the acid solution acts on the zinc, hydrogen gas is liberated. This gas is not only hot, but is apt to be laden with globules of acid. The function of the water in the wash-bottle is to cool the gas and to clean it before it goes to the finger pipe. Air pressure is pumped up into the tank. The petcock N in the branch pipe is then opened and the hydrogen gas issuing from the burner is ignited. The air is then admitted by opening the petcock S and adjusted until a hot-pointed flame of a greenish color is obtained that is suitable for burning. If any of the solution is spilled its

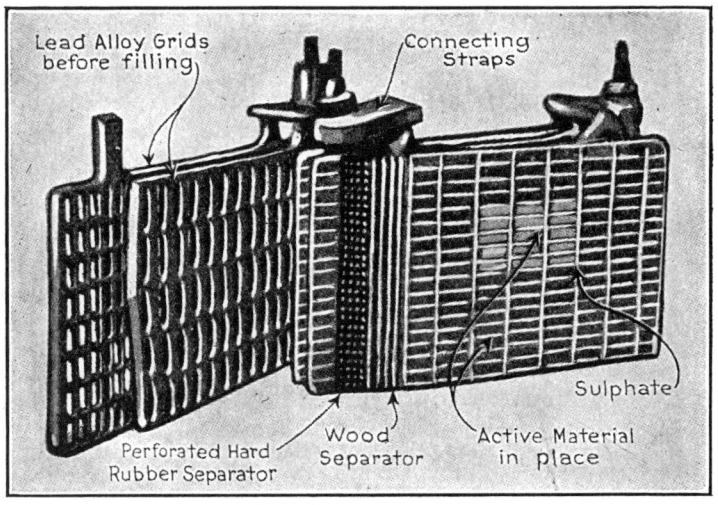

Fig. 35.—Showing Antimony-Lead Alloy Grids Before Filling and How Plates are Joined Together by Connecting Straps.

action may be neutralized by using Gold Dust, Pearline, washing soda, slaked lime, or ammonia. If burned when pouring acid into the reservoir, which, of course, will not occur unless this is carelessly done, apply olive oil to the burn and not water.

Another outfit suitable for lead burning consists of apparatus for burning a mixture of oxygen and illuminating gases. As oxygen is widely used in many garages for carbon removal, the same tank may be easily connected up to a single blow-pipe outfit. The connections are very clearly shown at Fig. 34. As the oxygen is carried under very high pressure in the tank it is necessary to provide a pressure reducer so that its pressure will not be too high at the burner. The illuminating gas is turned on first and ignited, after which the oxygen supply is regulated, so that a good-burning flame is secured.

Lead Burning.—Lead burning consists in melting the metals and causing the parts to flow together and become joined without the aid of solder. It requires considerably more skill than any other form of brazing or soldering. A long step toward success may be taken by the proper arrangement of the work. It is usual to provide something which may serve as a mould or guide for the melted metal. For example, if two lead sheets are to be united by soldering, they are laid on a sheet of some non-heat-conducting substance, such as brick or asbestos. The work in the immediate neighborhood of the joint is carefully scraped so as to remove all oxide or scale which would tend to bind the melted lead and prevent it from flowing freely. The metal at the seam is heated by a very hot bit or the flame from a blowpipe, so that there is a uniform flow of lead across the seam. It is sometimes necessary to add more lead to the seam by melting a strip held in the hand. A flame of some sort is the most satisfactory source of heat for the average lead-burning job, because not only is the heat more uniform, but also more intense, and the lead melts at the desired point before the surrounding metal becomes sufficiently hot to soften. There are several types of blowpipe for this purpose on the market. The flame is usually small, sharp-pointed, and very intense. Lead burning is absolutely necessary, and is insisted upon in certain classes of work, for instance, in lining tanks with lead for chemical solutions, or for joining the grids and lugs of storage batteries.

Directions for Lead Burning: To connect the various plates comprising an element, the Gould Storage Battery Company advise the use of a special fixture to insure accuracy in spacing. Be careful to select the proper spacer (Fig. 36), and attach it to the burning-rack. Place the plates on the burning-rack so that the lugs extend through the slots in the spacer. Fit the connecting strap over the lugs. Adjust the spacer by the adjusting nuts until the strap is at the proper height on the lugs. Using the flame, melt the lug to be burned and the adjoining material until they tend to run together. Using a piece of burning strip, melt the end thereof and fill in around the lug until the whole is a molten mass. Allow the joint to cool and cut off the protruding end of the lug with a pair of end-cutting pliers. Melt the remaining end of the lug till it flows into the strap. Repeat until all plates are burned to the strap.

Burning Plates to Old Straps: The storage battery company

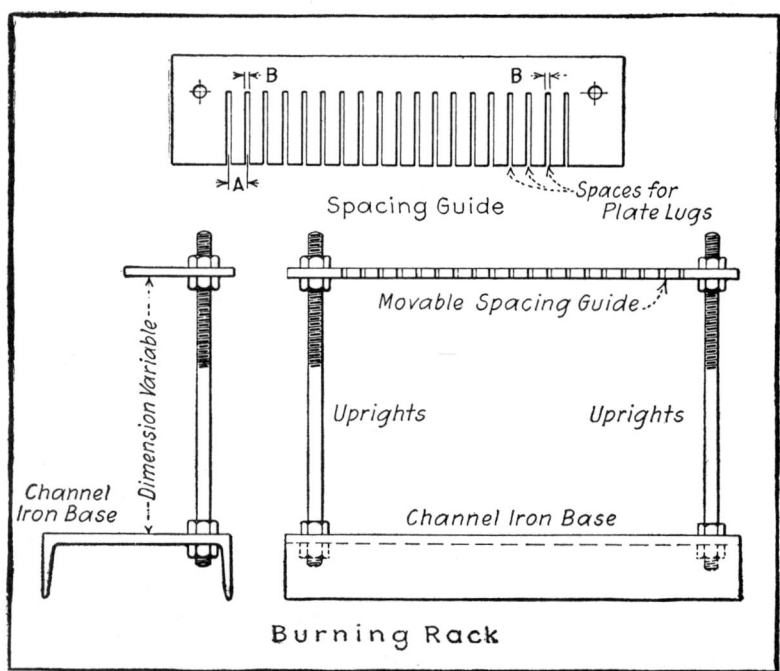

Fig. 36.—Gould Burning-Rack for Supporting Plates When Burning Plate Lugs to Busbars.

furnishes connector straps for nearly all the modern types of batteries. Sometimes it happens, however, that a battery of an old model or of a manufacture seldom used will be set up with straps that cannot be duplicated. Under these circumstances the old straps should be utilized. With the hacksaw cut off the plates.

LEAD BURNING

Cut slots in the strap, using the old lugs as a guide. Cleanse the strap thoroughly in ammoniated water and scrape clean. Using the proper spacer, proceed as described above.

To Burn Terminal Connector to Pillar Post: Scrape the parts clean. Fit the terminal connector to the pillar post. If the terminal connector does not set low enough, ream the terminal with the triangular scraper until the fit is exact. Heat with flame until the inside of terminal connector and outside of pillar post are one molten mass—*throughout*. Fill in with molten, burning material and allow to cool. To burn connecting link to pillar posts, proceed as above.

STORAGE BATTERIES SIMPLIFIED

AUTOMOBILE LIGHTING AND STARTING.
BATTERY DEFECTS AND RESTORATION SUMMARIZED.

Defect	Symptom	Remedy
Broken or cracked cell jar.	Low level in one cell.	Replace with new jar.
Plates sulphated. Active material crystallized.	Gravity will not rise on charge.	Long slow charge at min. rate.
Lack of acid.	Gravity will not rise on charge.	Mix new electrolyte and fill cells recharged.
Electrolyte low.	Overheating.	Refill with water or electrolyte, depending on gravity.
Rapid charging.	Overheating.	Regulate generator output.
Solution level too high.	Electrolyte leaks out of vents.	Draw out surplus with syringe.
Cracked cell cover. Defective sealing. Poor vent.	Battery box eaten. Terminals corroded.	Do not fill cells so much.
Undercharging.	Battery capacity low.	Augment generator output; charge battery from outside source.
Charging too fast. Buckled or warped plates.	Overheating.	Charge at lower rate. Cell temperature must not be above 100° F.
Short circuits. Large sediment deposit.	Battery loses charge rapidly when idle.	Go over external wiring. Clean out sediment.
No current in cold weather.	Battery frozen.	Cannot usually be repaired. Try slow long charge.
Rotting wood box. Verdigris on terminals. Rotting conductor wire insulation.	Too much acid or electrolyte.	Take out surplus.
Separator failure.	Impure water. Electrolyte too rich in acid.	Use only distilled water. Dilute rich electrolyte.
Separators charred or punctured.	Overheating. Loss of charge.	Replace separators. Maintain level of electrolyte.
Lights uncertain.	Battery nearly discharged.	Give boosting charge from outside source.
Current output low even though liquid is at proper level.	Gravity of electrolyte too low.	Bring gravity up to 1.280° by charging.
Excessive current consumption.	Gravity of electrolyte down to 1.100°.	Give long slow charge 3 to 5 amps. rate.
One cell defective. Poor separators.	Total voltage low. Weak current.	Rebuild poor cell.
Active material shedding.	Large sediment deposit.	Rebuild battery.
Battery not properly fastened down.	Cell jars break or crack.	Fit proper hold-down clamps.
Acid escapes through vents.	Terminals corroded.	Clean with ammonia or washing soda; coat with vaseline.
Excessive gassing. Poor box ventilation.	Metal battery box corroded.	Use lower charging rate-coat box interior with asphaltum paint.
Battery discharged.	Starting motor will not start engine. Lights burn dim.	Give thorough charge from outside current.
Generator not charging properly.	Battery needs frequent boosting charges.	Overhaul generator. Regulate for proper charging rate.

CHAPTER IV

Battery-Charging Methods—Currents and Voltages—Electrolytic Rectifiers—Vibrator Rectifiers—Mercury Arc Rectifiers—Rotary Converters—Rheostats—Lamp-Bank Resistance—Charging Precautions—Charging Vehicle Batteries—Winter Care of Automobile Storage Batteries.

THE equipment to be used in charging storage batteries depends entirely upon the type and size of batteries to be charged, the current voltage and character available for charging, and the individual characteristics of the batteries themselves. Storage batteries can be charged only with direct current, i. e., that which flows always from the same direction. It is evident that the use of alternating current, if the mains were attached directly to the battery, would result in rapid changes in the interior of the cells, and as the flow in one direction would tend to neutralize that in the other, the plates would depreciate very rapidly. If alternating current is the only kind available, this must be transformed or rectified into direct current. All cells cannot be charged at the same rate. The greater the capacity of the battery and the higher its discharge rate, the greater the amperage of the current that can be used in charging. While the voltage of a storage battery made of certain materials will not vary with the size, the amperage or current output increases with the plate size and number. A lead-plate storage battery no longer than a thimble will have just as high voltage as one as big as a barrel. It will be evident, however, that if too much current is passed through a small cell it will be injured, whereas too little current passed through a large cell will have but little effect on changing the character of the plates.

There are two general methods in use for charging either the sulphuric-acid-lead batteries or the alkali-nickel-iron batteries used in the various commercial applications. The first method, and the one most widely followed, is called the constant-current system. The other system, which has only received attention lately, is called the constant-potential method. Two other schemes are used also which are modifications of the two previously named general

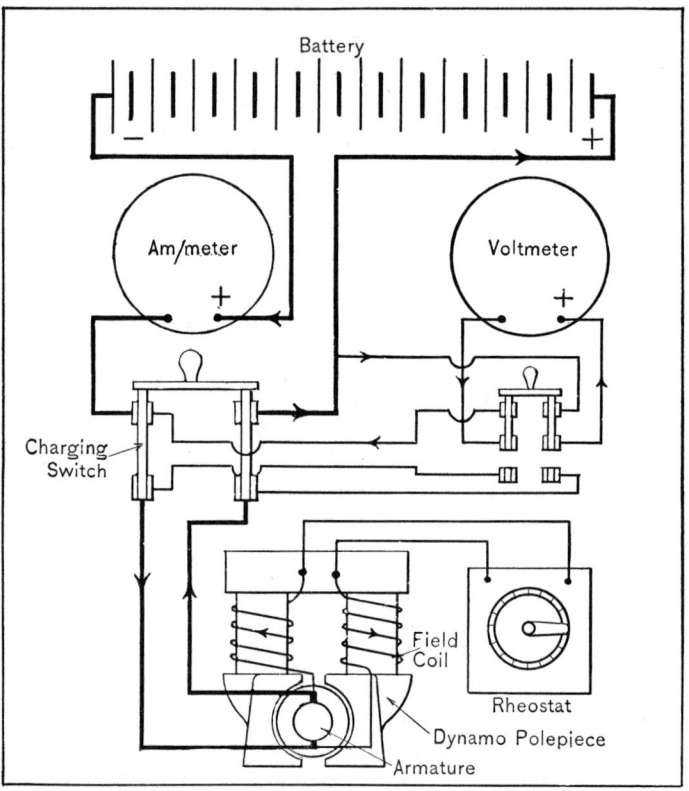

Fig. 37.—Battery-Charging Outfit, Having Field-Coil Rheostat.

methods, one known as the multiple voltage system, the other as a fixed resistance method. The constant potential method is said to offer a number of advantages. There is less evaporation of electrolyte and less shifting of rheostats is needed. Any form of battery may be charged by means of a fixed resistance connected in series. When the battery counter electromotive force is nearly

BATTERY-CHARGING METHODS 93

equal to the voltage of the line, this method approximates the general characteristics of the constant potential system. If the voltage of the battery is considerably less than that of the charging current, the characteristics will approximate the constant-current charging method. It may be stated that the variation in the current used for charging is inversely proportional to the difference between the maximum counter voltage of the battery and the potential or voltage of the supply circuit.

The usual method of charging batteries in garages is to con-

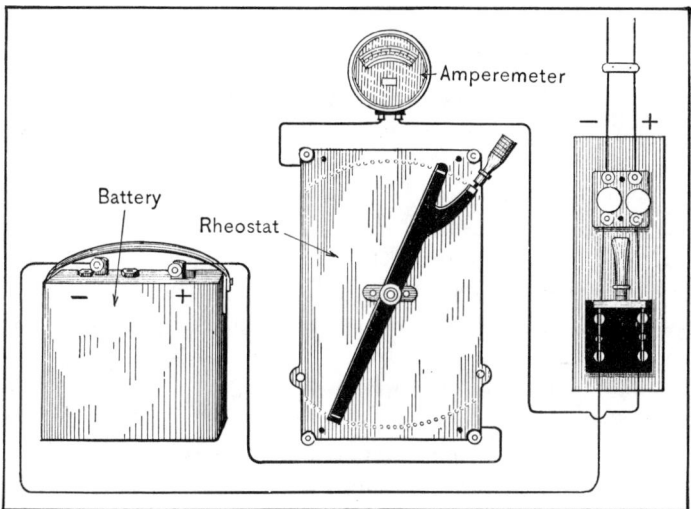

Fig. 38.—Direct-Current Battery-Charging Outfit, With Rheostat in Line Between Battery and Main-Line Switch.

nect them up in series. In order to meet the requirements of the best battery-charging practice it is stated that they should never be charged in series unless they are all composed of the same general type, size and capacity of cells, and that all of the batteries are in the same state or condition of discharge. The disadvantages of this method are that the batteries that have not been discharged so much as the others are apt to be overcharged, and if the battery capacity varies very much, the charging rate may be too low for some cells and too high for others. These disadvantages do not apply in charging vehicle batteries where the individual cells are of the same size, type and capacity. The only way the series charging can be carried on is by using a compromise charging rate and carefully testing the various batteries from time to time to make sure that they will be removed when properly charged. The types and sizes of cells used in automobile starting, lighting and ignition batteries do not vary as much as might be expected, and if a compromise charging rate is intelligently selected, it is a method that gives fairly good results in practice, though it is theoretically wrong.

When a direct current of 110 volts potential is available Edison batteries composed of 60 cells or lead-plate batteries of 40 to 44 cells may be charged directly from the line by means of a rheostat to regulate the amount of current passing through the batteries, which is placed in series with each battery. If the service is of higher voltage, a motor generator or rotary converter set may be used. If alternating current only is available, various types of rectifiers are needed. If only one battery is to be charged and if the voltage of the generator can be adjusted by means of a rheostat connected in series with the field coils of the dynamo, as shown at Fig. 37, then no rheostat will be needed between the battery and the dynamo, because the charging current can be kept to the proper value by varying the dynamo voltage. When a mercury arc rectifier is employed in battery charging, the charging current can be regulated by control dials and a rheostat is not needed. The various types of rectifiers suitable for use with alternating current are to be described in proper sequence. Rotary converters may be used with either direct or alternating current, depending entirely upon the type of motor used. It is evident that the dynamo of such a combination must always be of the direct-current type, though its output will vary according to the power of the actuating motor, the method of winding, and wire used in field and armature coils. The motor may be a direct-current type, wound for high voltage, or it may be a type wound to operate on alternating current.

The arrangement of the essential parts of a typical battery-charging system where the current value is altered by a dynamo

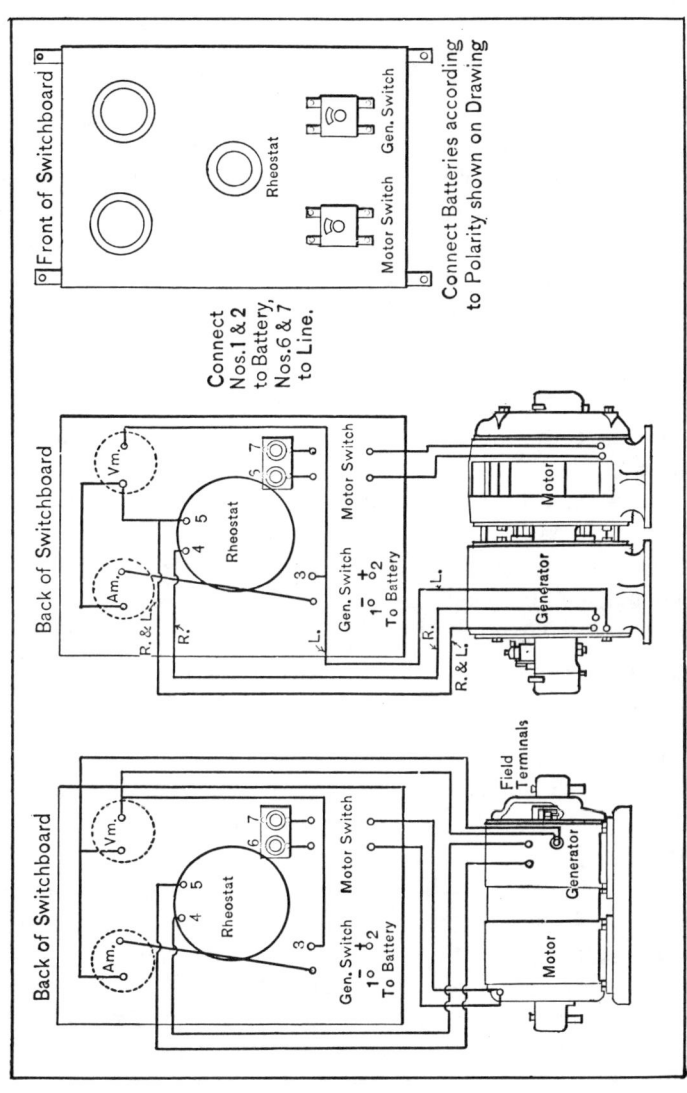

Fig. 39.—How to Wire Rotary Converter Charging Outfit, With Ammeter Rheostat and Voltmeter, Showing Front and Rear of Switchboard.

field-coil rheostat is shown at Fig. 37. A shunt-wound generator is employed, and the main leads from the armature brushes are connected to the lower poles of a double-pole knife-switch. The battery is connected to the upper portion of the switch, an amperemeter being placed in circuit as indicated. The hinges of a small double-throw, double-pole switch are connected with a voltmeter. The amperemeter and voltmeter should be of the permanent type. Before throwing in the charging switch it is possible to read the voltage of the battery, and also by throwing the switch to read that of the charging generator so that it may be adjusted to a slightly greater voltage. The main switch is then closed and the rheostat used to raise the voltage sufficiently to drive a suitable charging current through the battery. With a system of this kind a circuit breaker or automatic overload switch should be included in the main line to protect the apparatus in case of accidental short circuit. An underload circuit breaker should also be provided to shut off the battery if the current falls to such a point that the battery will discharge through the generator. These are not shown in the simplified wiring diagram, neither are the fuses that prudence dictates should be used.

The Westinghouse Vibrator Rectifier is an inexpensive form of apparatus to charge small batteries from ordinary lighting circuits. The device, which is shown at Fig. 42 A, reduces the voltage of the lighting circuit to the proper value by the use of a small double step-down transformer, and rectifies this reduced voltage to the uni-directional voltage necessary for battery charging by electrically operating switching mechanism. The transformer serves the double purpose of diminishing the line voltage for the battery to be charged and also for providing a return path for the direct current. The charging current flows from one end of the secondary winding, and after passing through a regulating resistance passes through a pair of contacts, which are closed automatically and at the proper time, and out from the center point of the armature to the battery, from which it returns to the neutral point of the transformer. During the next half cycle the voltage in the transformer secondary is reversed in direction and the other pair of contacts is closed and the voltage is applied to

BATTERY-CHARGING METHODS

the battery from the half of the secondary that has previously been idle. As the current flow is in the same direction as that previously supplied, the battery is charged exactly the same as if uni-directional current from a generator was used.

The element upon which the success of the outfit depends is the vibrating mechanism, upon which devolves the duty to reverse connections in synchronism with the voltage and also exactly in step with the transformer secondary voltage in such a manner as to open the current character circuit at the instant of zero current and prevent injurious wear of the contacts by sparking. The following description of the action of this rectifier is repro-

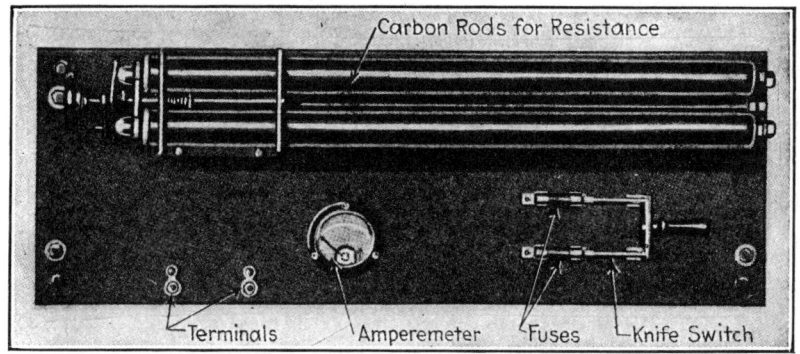

Fig. 40.—Carbon Rod Rheostat.

duced from the *Electric Journal,* and the action of the device may be understood by studying the wiring diagram shown at Fig. 42 B. "Two small laminated iron magnets, marked *A.C.* magnets, are connected in series across one-half of the transformer secondary, connections being made so that the corresponding ends of the magnet are of the same magnetic polarity. A direct-current magnet, polarized by shunt current from the battery, is so placed as to bring its ends within the effective field areas of the *A.C.* magnets. Since the ends of the *D.C.* magnet are of opposite polarity, they are forced at any instant in opposite directions by the fields of the *A.C.* magnets and one pair of contacts is closed. During the succeeding half cycle the *A.C.* magnets are reversed in polarity, while the *D.C.* magnet is not; the impelling force is reversed, and the armature takes such a position as to close the other pair of contacts. One side of the battery is thus connected alternately to the opposite ends of the secondary of the transformer in synchronism with the alternating voltage, while the other side is permanently connected to the center point. Exact timing to insure sparkless operation, by

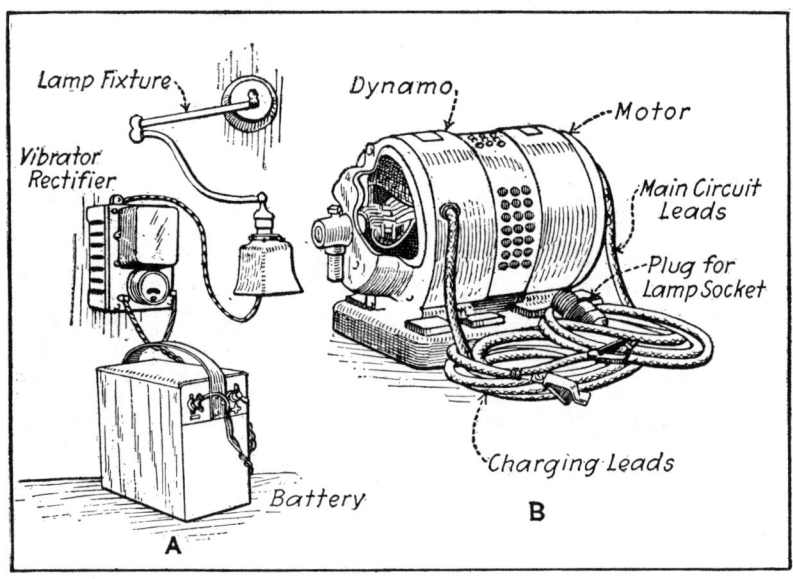

Fig. 41.—Devices for Rectifying Alternating Current. A—Vibrator Type Rectifier. B—Small Rotary Converter Set.

breaking the current-carrying circuit at the time when the battery and transformer voltages are equal and opposite and no current is flowing, is secured by connecting in series with the *A.C.* magnets, a resistance which alters the power-factor of the current in the magnets without affecting that of the load current in the transformer. This change in power-factor translates in time the impelling force, with respect to the current in the contacts, and secures the result of sparkless operation. This phase-controlling resistance is made variable, in order that the outfit will be applicable on circuits of which the wave form is not a true sine wave, and on circuits on which the voltage is not of normal value. The condensers connected around the contacts reduce to a negligible amount the unavoidable slight sparking, due to fluctuations in the line voltage, variation in wave form and change in battery voltage.

"The regulating resistance, which is connected in each side of the secondary circuit between the transformer and the stationary contact, is for the purpose of giving the outfit high or low regulation, in order that the change in battery voltage, as the charge progresses, will make only a small change in the current delivered. The standardization of lighting batteries in general use has resulted in the selection for the commercial form of this apparatus of such transformer voltage and resistance value as to make the charging current under normal conditions approximately 8.5 amperes at the start of charge and 6.5 amperes at the finish. The features above mentioned result in an outfit which can be connected to an ordinary alternating-current lighting circuit and to a battery, without attention to polarity, owing to the polarization of the $D.C.$ magnet by the battery, and which will then, after a single adjustment of the phase-controlling resistance, give a full charge to the ordinary lighting battery without further attention. The cost of power for such a charge at the common rate of 10 cents per kilowatt hour is roughly 6.5 cents, as compared to the ordinary charge of 75 cents to $1.25 per charge by a public garage."

Battery-Charging Apparatus.—The apparatus to be used in charging a storage battery depends upon the voltage and character of the current available for that purpose. Where direct current can be obtained the apparatus needed is very simple, consisting merely of some form of resistance device to regulate the amperage of the current allowed to flow through the battery. The internal resistance of a storage battery is very low, and if it were coupled directly into a circuit without the interposition of additional resistance, an excessive amount of current would flow through the battery and injure the plates. When an alternating current is used it is necessary to change this to a uni-directional flow before

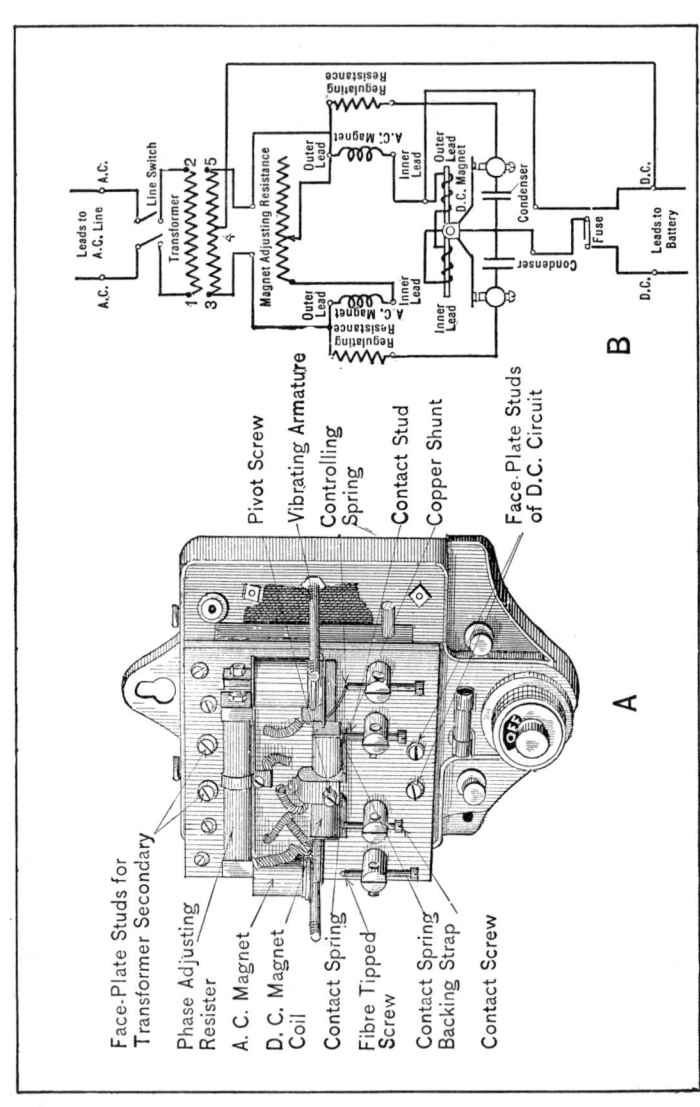

Fig. 42.—Construction of Westinghouse Vibrator Type Alternating-Current Rectifier at A, and Wiring Diagram Showing Operation at B.

it can be passed through the battery. Alternating current is that which flows first in one direction and immediately afterward in the reverse direction. When used in charging storage batteries some form of rectifier is essential. The rectifier may be a simple form, as shown at Fig. 43 A, which is intended to be coupled directly into a lighting circuit by screwing the plug attached to the flexible cord in the lamp socket. A rotary converter set, such as shown at B, or at Figs. 39 and 41, may also be used; in this the alternating current is depended on to run an electric motor, which drives the armature of a direct-current dynamo. The current to charge the battery is taken from the dynamo as it is suitable for the purpose, whereas that flowing through the motor cannot be used directly.

The view at Fig. 43 C shows a usual form of hydrometer syringe which is introduced into the vent hole of the storage battery, such as shown at Fig. 44, and enough electrolyte drawn out of the cell to determine its specific gravity. This is shown on the hydrometer scale, as indicated in the enlarged sections. A very useful appliance where considerable storage-battery work is done is shown at Fig. 45 A. This is a stand of simple form, designed to carry a carboy containing either acid, distilled water or electrolyte. In fact, it might be desirable to have three of these stands, which are inexpensive, one for each of the liquids mentioned. In many repair shops the replenishing of storage batteries is done in a wasteful manner, as the liquid is carried around in a bottle or old water pitcher and poured from that container into the battery, often without the use of a funnel. The chances of spilling are, of course, greater than if the liquids were carefully handled and more time than necessary is consumed in doing the work. The stand shown is about 5 feet high and is fitted with castors so it may be easily moved about the shop if necessary. For example, in taking care of electric vehicle batteries, it may be easier to move the carboy to the battery than to remove the heavy battery from the automobile. The container for the liquid is placed on top of the stand and the liquid is conveyed from it by a rubber tube. The rubber tube is attached to a glass tube extending down nearly to the bottom of the liquid. At the bottom

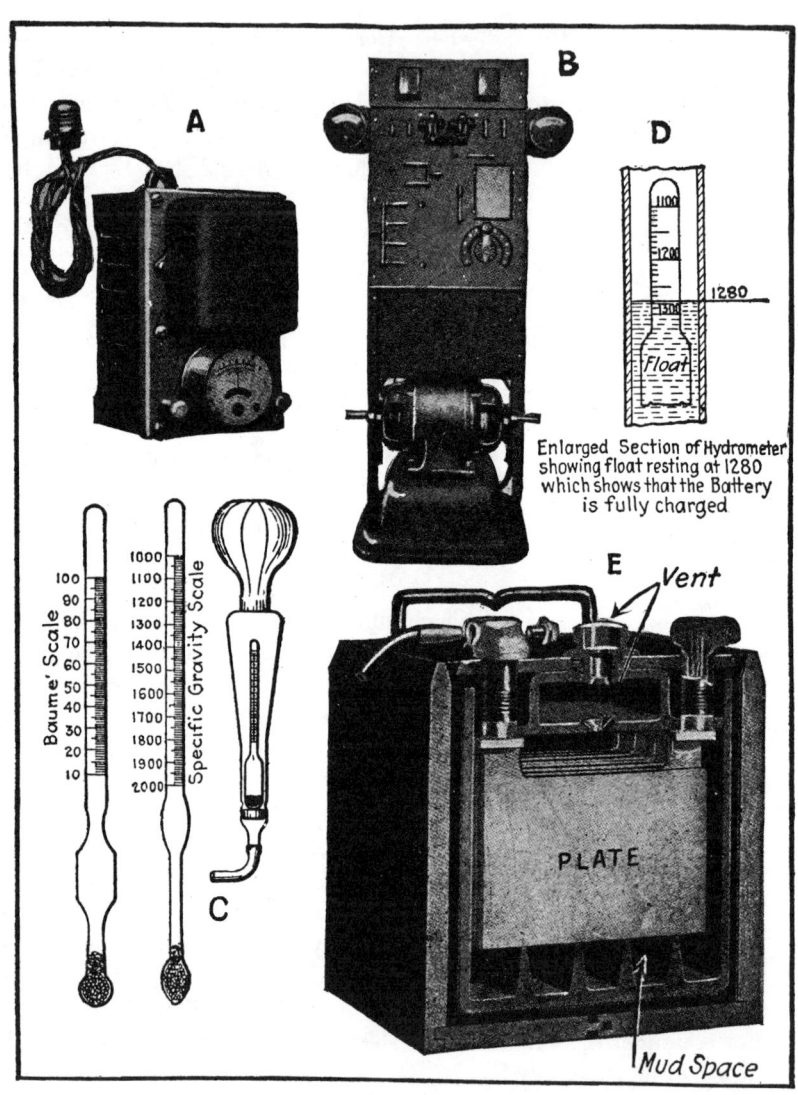

Fig. 43.—Devices Used in Charging and Caring for Storage Batteries.

BATTERY-CHARGING METHODS 103

of the rubber tube an ordinary chemist's clip, which controls the flow of liquid, is placed. In order to start a flow of liquid it is necessary to blow into a bent glass vent tube, which is also inserted into the stopper. Once the rubber tube has become filled

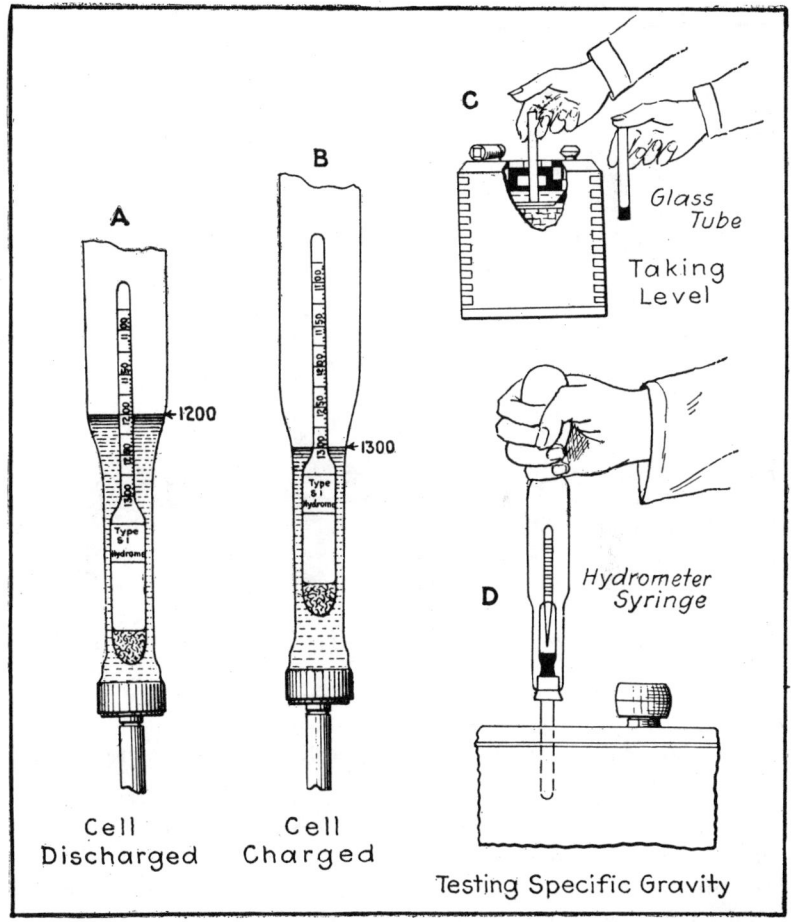

Fig. 44.—Outlining Positions of Hydrometer in Electrolyte When Cell is Discharged at A and When Cell is Charged at B. Method of Determining Electrolyte Level at C. How to Take Specific Gravity Reading Shown at B.

104   STORAGE BATTERIES SIMPLIFIED

with liquid, merely opening the clip will allow the liquid to flow into the battery as desired.

In most communities the incandescent lighting circuit is used for charging batteries on account of the voltage of the power circuits being too high. The incandescent lighting circuit may be

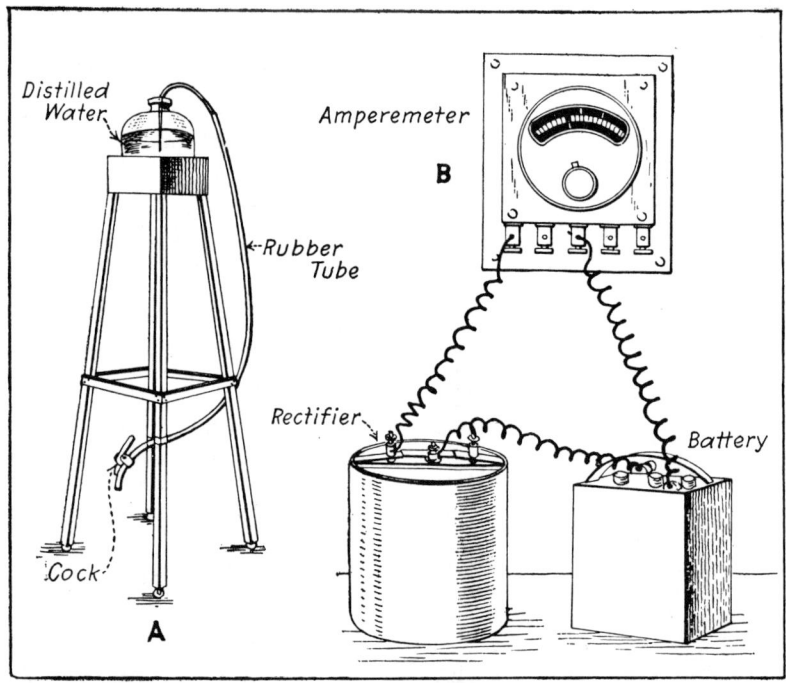

Fig. 45.—Simple Stand for Carrying Electrolyte or Distilled Water Bottle at A. Method of Using Rollinson Electrolyte Rectifier Shown at B.

any one of six forms. A direct current of either 110 or 220 volts used over short distances, either 220 or 440 volts on three-wire circuits over long distances, alternating current at a constant potential, usually 110 volts and in various polyphase systems. It might be stated that in the majority of instances house and garage lighting circuits furnish direct current of 110 volts. We will consider the devices used with the alternating form, one of which is shown at Fig. 45 B. This is known as the Rollinson electrolytic rectifier, which is based upon the following principles: When an element of aluminum and a corresponding element or plate of iron are submerged in a solution of certain salts, using these elements as negative and positive terminals, respectively, the passage of an

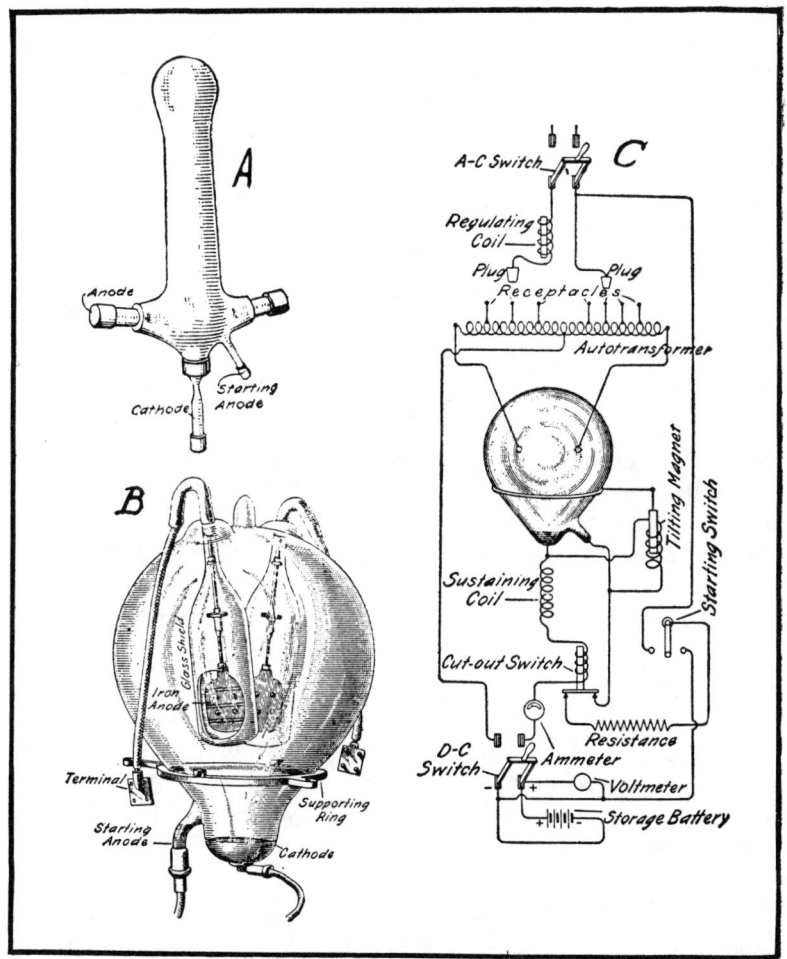

Fig. 46.—Mercury Rectifier Bulbs and Methods of Wiring to Charge Storage Battery From Alternating-Current Mains.

106 STORAGE BATTERIES SIMPLIFIED

electric current through the solution produces a chemical action which forms hydroxide of aluminum. A film of hydroxide thus formed on the aluminum element repels the current. The arrangement of the cell will then permit current to pass through it in one direction only, the film of chemical preventing it from passing in the opposite direction. The result is that if an alternating current is supplied to the cell a direct pulsating current can be obtained from it. The outfits usually include a transformer for reducing the line voltage to the lower voltages needed for battery-charging purposes. Regulation of the current is effected in the simplest type by immersing the elements more or less in the solution in the jar. As complete instructions are furnished by the manufacturers, it will not be necessary to consider this form of rectifier in detail.

One of the most commonly used rectifying means is the mercury arc bulb. This device is a large glass tube of peculiar shape, as shown at Fig. 46, which contains a quantity of mercury in the base. On either side of this lower portion two arms of the glass bulbs extend outwardly, these being formed at their extremities into graphite terminals or anodes, indicated as A and A-1 in the diagram at Fig. 47. The current from the auto transformer is then attached one to each side. The base forms the cathode or mercury terminal for the negative wires. The theory of this action is somewhat complicated, but may be explained simply without going too much into detail. The interior of the tube is in a condition of partial vacuum, and while the mercury is in a state of excitation a vapor is supplied. This condition can be kept up only as long as there is a current flowing toward the negative. If the direction of the current be reversed so that the formerly negative pole becomes a positive the current ceases to flow, as in order to pass in the opposite direction it would require the formation of a new cathode element. Therefore the flow is always toward one electrode, which is kept excited by it. A tube of this nature would cease to operate on alternating-current voltage after half a cycle if some means were not provided to maintain a flow continuously toward the negative electrode. In the General Electric rectifier tube there are two anodes and one cathode. Each of the former is connected to a separate side of the alternating current supply and also through reactances to one side of the load and the cathode to the other. As the current alternates, first one anode and then the other becomes positive, and there is a continuous flow toward the mercury cathode, thence

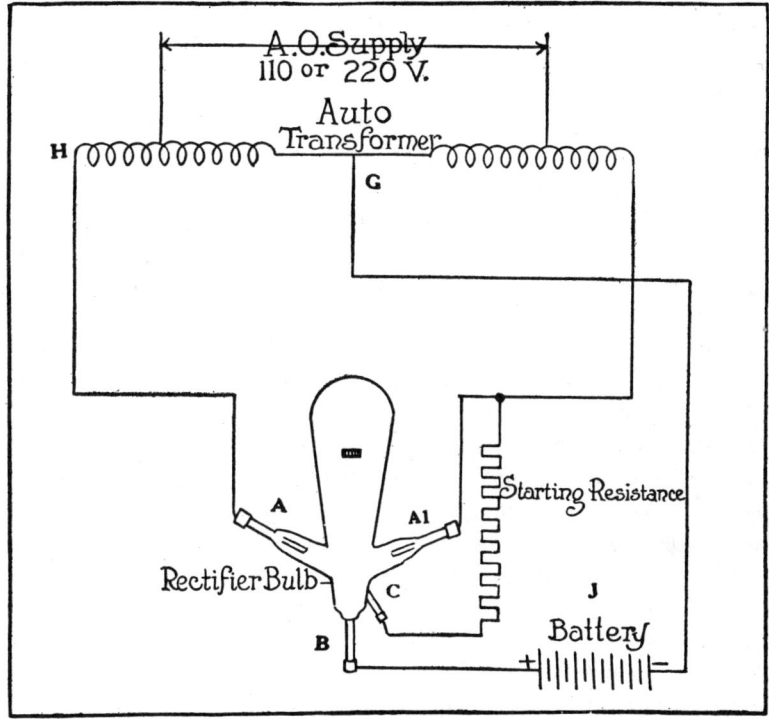

Fig. 47.—Wiring Diagram Defining Use of Mercury Arc Rectifiers.

through the load (in this case the battery to be charged), and back to the opposite side of the supply through a reactance. At each reversal the latter discharges, thus maintaining the arc until the voltage reaches the value required to maintain the current against the counter E. M. F. and also reducing the fluctuations in the direct current. In this way a true continuous flow is obtained, with very small loss in transformation.

A small electrode connected to one side of the alternating circuit is used for starting the arc. A slight tilting of the tube makes a mercury bridge between the terminal and draws an arc as soon as the tube is turned to a vertical position. The ordinary form used for vehicle batteries has a maximum current capacity of 30 amperes for charging the lead plate type, and a larger form, intended for use with Edison batteries, yields up to a limit of 50 amperes. Those for charging ignition batteries will pass 5 amperes for one to charge six cells and a larger one that will pass 10 amperes for from three to ten batteries. As is true of the electrolytic rectifier, complete instructions are furnished by the manufacturer for their use.

The Wagner device, which is shown at Fig. 43 A, operates on a new principle, and comprises a small two-coil transformer to reduce the line voltage to a low figure, the rectifier proper, which consists of a vibrating armature in connection with an electromagnet, and a resistance to limit the flow of the charging current. A meter is included as an integral part of the set for measuring the current flow. All sets are sold for use with ignition or lighting batteries of low voltage, with a lamp socket-plug and attaching cord, the idea being to utilize an ordinary lighting circuit of 110 volts A. C. The magnet and vibrating armature accomplish the rectification of the current with little loss, the action after connection to the battery which is to be charged proceeding automatically. By a simple device, the current stoppage throws the main contacts open, so the partially charged battery cannot be rapidly discharged. While the rectifiers are constructed to use 60-cycle, 110-volt alternating current, they will work at all frequencies from 57 to 63. The size made will pass three to five amperes, the voltage being sufficient to recharge a three-cell battery.

When batteries are to be charged from a direct current it is possible to use a rheostat to regulate the voltage at the terminals. The construction of a rheostat is very simple, as it consists only of a group of high-resistance coils of wire mounted in insulating material, and having suitable connections with segments on the base plate, upon which is mounted the operating arm that makes the contact. According to the manner in which these are made

BATTERY-CHARGING METHODS

and wired a large resistance is introduced at first, gradually decreasing as the lever is moved over, or it may operate in the reverse fashion, a large amount of current being allowed to pass at the first contact and less as the handle progresses across the path. Rheostats should only be purchased after consulting a capable electrician, as the required resistance must be figured out from the voltage of the circuit to be used, the maximum battery

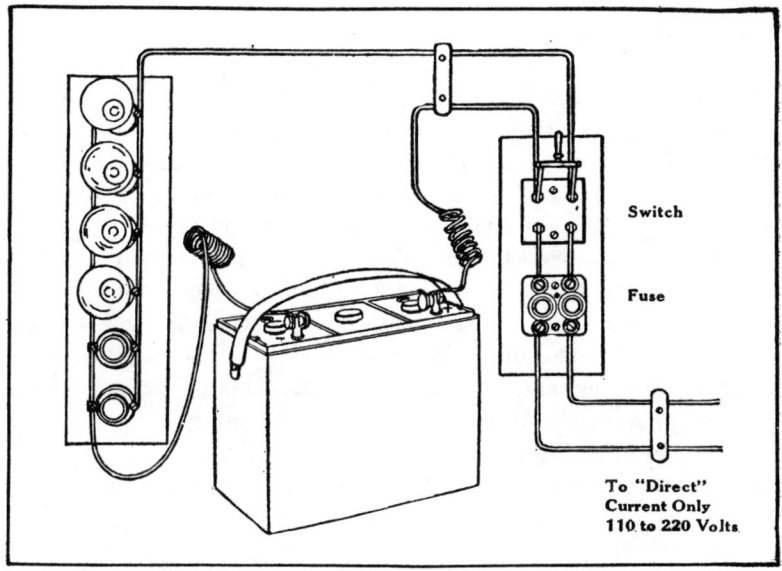

Fig. 48.—Charging Storage Battery From Direct Current With the Lamp-Bank Regulation.

current, the charging rate in amperes and the number of cells to be charged at one time.

By far the simplest method of charging storage batteries is by interposing a lamp-bank resistance instead of the rheostat. These are easily made by any garage mechanic and are very satisfactory for charging ignition or lighting batteries. Standard carbon lamps of the voltage of the circuit shown should be used, and the amperes needed for charging can be controlled by varying the candle power and the number of lamps used. If the lamps are to operate on 110-volt circuit, a 16-candle-power carbon filament lamp will permit one-half ampere to pass; to 32-candle-power will allow 1 ampere to pass. If it is desired, therefore, to pass three amperes through the battery, one could use 3 32-candle-power lamps, or 6 16-candle-power lamps. If the lamps are to burn on 220 volts, it should be remembered that when the voltage is doubled the amperage is cut in half, therefore the 32-candle-power, 220-volt carbon filament bulbs will only pass half an ampere. The method of wiring is very simple, as may be readily ascertained by referring to Fig. 48. The line wires are attached to a fuse block and then to a double knife switch. The switch and fuse block are usually mounted on a panel of insulating material such as slate or marble. One of the wires, the positive of the circuit, runs from the switch directly to the positive terminal of the storage battery. The negative wire from the switch passes to the lamp-bank resistance. The lamps are placed in parallel connection with respect to each other, but in series connection in respect to the battery. When coupled in this manner the current must overcome the combined resistance of the storage battery, which is very low, and that of the lamps. This prevents the battery being charged with current of too high voltage.

A water resistance is easily constructed by using a small wooden tub or half barrel. Two sheet-lead plates are suspended from wood sticks resting on top of the tub, the supports being movable to bring the lead plates closer together or separate them, as desired. A wire is brought from the battery, as shown in Fig. 49, to one side of the switch, then to one of the plates in the water resistance, then from the other plate of the water resistance to an ammeter; to the other side of the switch and from there to the opposite pole of the battery. Such a resistance is used for making a test discharge of a vehicle battery; it would not be a very practical way of charging batteries because of the great absorption of current by the water. Before starting a discharge, care should be taken to have the lead plates and wires separated. The tub can then be filled with clean water and the switch closed. A small quantity of electrolyte should then be poured in the water, a very little at a time, until the ammeter shows that the proper

BATTERY-CHARGING APPARATUS 111

amount of current is flowing. The farther apart the plates are, the greater the resistance. As more electrolyte is added, even if the plates are not disturbed, the resistance becomes less. Never let the plates touch each other.

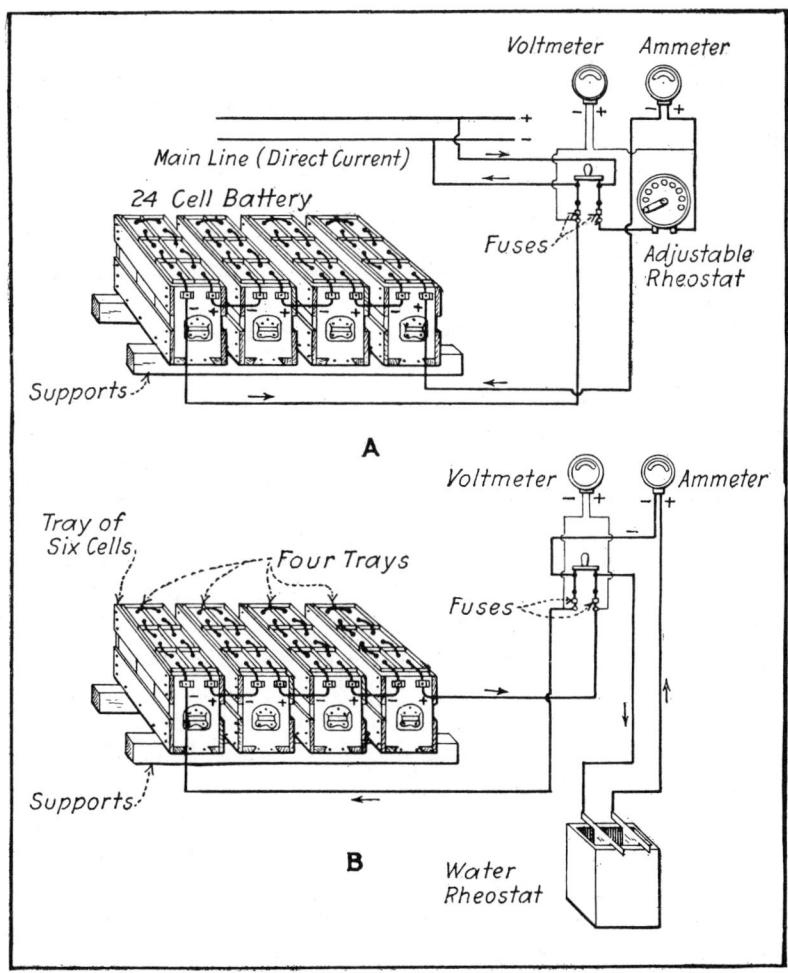

Fig. 49.—Method of Charging 24-Cell Vehicle Battery at A. How Water Rheostat is Used in Making Test Discharge Outlined at B.

The points to be especially emphasized in connection with the charge are:

First—On regular charges keep the rates as low as practical and cut off the current promptly. It is preferable to cut off a little too soon rather than to run too long where there is any question.

Second—Overcharges must be given at stated intervals and continued to a complete maximum. They should be cut off at the proper point, but when in doubt it is safer to run too long, rather than to cut off too soon.

Third—Do not limit the charge by fixed voltage.

Fourth—Keep the temperature within safe limits.

Fifth—Keep naked flames away from cells while charging, as the gas given off is inflammable. Always see that gas vents are clear before charging.

Winter Care of Storage Batteries.—It would not do simply to leave the battery in the car for a period of, say, four or five months without giving it any care or attention, for in that case at the end of that time it would be found to have its plates so thickly covered with lead sulphate as to make it practically useless. For storage batteries "to rest is to rust" and become ruined, unless special precautions are taken. Automobile storage batteries are all or nearly all of the sealed-in type, from which the elements cannot be removed without a great deal of trouble. Therefore, the only method of keeping the plates intact consists in charging the battery at intervals of about two weeks. The following advice concerning the care of batteries during a protracted period of idleness of the car is due to the Willard Storage Battery Company, and refers especially to the batteries of starting and lighting systems.

At intervals of two weeks the engine should be run until the electrolyte shows a specific gravity of 1.280. If this is done regularly the engine need be run only about an hour each time. But if the owner should not be in possession of an hydrometer, it is better to run the engine two or three hours each time, for the sake of safety. To charge the battery properly the engine should be run at a speed corresponding to a car speed of about 20 mph

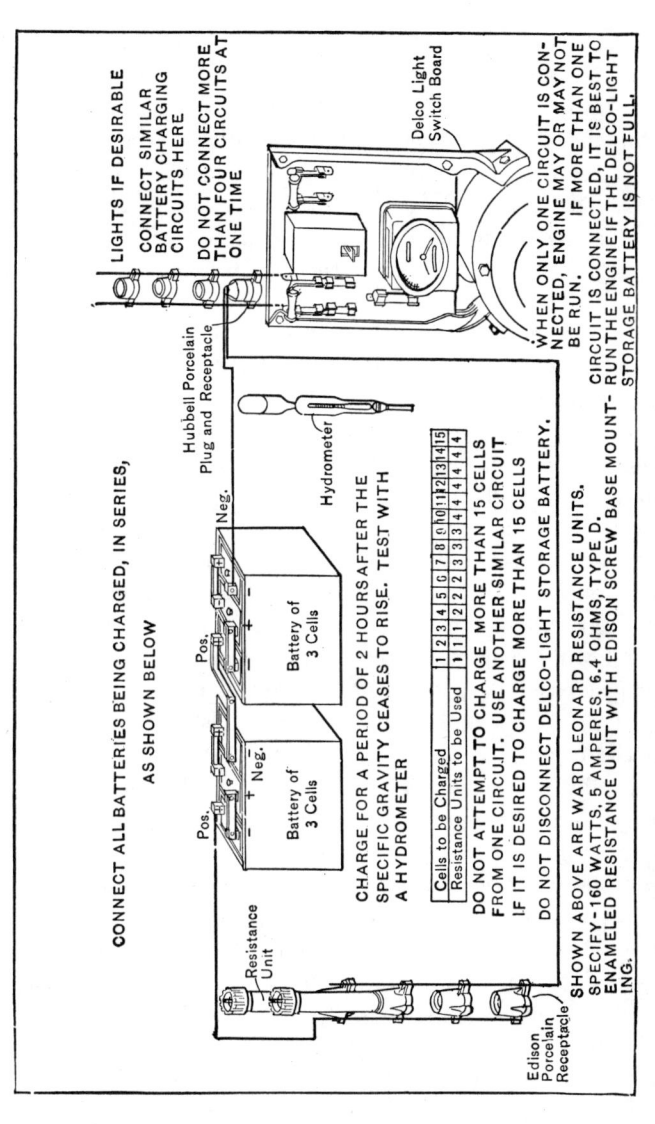

Fig. 50.—Diagram Outlining Necessary Connections for Using Delco Battery Charging Equipment.

on the direct drive. There may be cases, however, where the owner is compelled to store his car in a space where it is practically impossible to run the engine. Where this is the case, it is recommended, if electric current is available, that the owner purchase a rectifier or small charging machine. A charge over night, or for about twelve hours, every two weeks with this apparatus, will be sufficient to keep the battery in a healthy condition. Before beginning the charging the battery should be inspected to see if it is filled with solution. If the solution needs replenishing, distilled water should be added until the solution fully covers the plates, which may be determined by removing the vent plugs and looking down into the cells. In case it is impossible to run the engine for charging and the owner does not care to incur the expense of purchasing a rectifier, he should remove the battery from the car and arrange for its storage at a garage which has charging facilities, stipulating that it must be charged every two weeks. The cost of having it so cared for will be nominal and will prove excellent insurance against deterioration.

To care for storage batteries of a type that is easily taken apart the following method is recommended: First charge the battery until every cell is in a state of complete charge. If there should be any short-circuited cells they should be put into condition before the charge is commenced, so that they will receive the full benefit of the charge. Then remove the elements from the jars, separating the positive from the negative groups, and place in water for about one hour to dissolve out any electrolyte adhering to the plates. Then withdraw the groups and allow them to drain and dry. The positives when dry are ready to be put away. If the negatives in drying become hot enough to steam, they should be rinsed or sprinkled again with clean water and then allowed to dry thoroughly. When dry, the negatives should be replaced in the electrolyte (of from 1.275 to 1.300 specific gravity), care being taken to immerse them completely and allow them to soak for three or four hours. Two groups may be placed in a jar and the jar filled with electrolyte. After rinsing and drying the plates are ready to be put away.

The rubber separators should be rinsed in water. Wood sepa-

TABLE I—CHARGING RATES

Ampere-Hours Discharged	Time Available until next Adjustment of Charging Current.							
	¼ hour	½ hour	¾ hour	1 hour	1¼ hours	1½ hours	1¾ hours	2 hours
	Amperes	Amperes	Amperes	Amperes	Amperes	Amperes	Amperes	Amperes
10	8	6	5	5	4	4	3	3
20	16	13	11	10	9	8	7	6
30	24	20	17	15	13	12	11	10
40	32	26	23	20	18	16	14	13
50	40	33	28	25	22	20	18	16
60	48	40	34	30	26	24	22	20
70	56	46	40	35	31	28	25	23
80	64	53	45	40	35	32	29	27
90	72	60	51	45	40	36	33	30
100	80	66	57	50	44	40	36	33
110	88	73	63	55	49	44	40	37
120	96	80	68	60	53	48	43	40
130	104	87	74	65	58	52	47	43
140	112	93	80	70	62	56	51	47
150	120	100	86	75	67	60	54	50
160	128	106	91	80	71	64	58	53
170	136	113	97	85	75	68	62	57
180	144	120	103	90	80	72	65	60
190	152	127	108	95	84	76	69	63
200	160	133	114	100	89	80	73	67
210	168	140	120	105	93	84	76	70
220	176	147	126	110	98	88	80	73
230	184	153	131	115	102	92	84	77
240	192	160	137	120	106	96	87	80
250	200	167	143	125	111	100	91	83

EXPLANATION.—In the left-hand column find the figure nearest to the ampere-hours discharged from the battery; follow across to the column headed by the available time. The figure at this intersection is the current to be used.

EXAMPLE.—Ampere-hour meter reading, 103 ampere-hours discharged; time available for boosting, one hour. Start at 100 in the left-hand column; follow across to the column headed 1 hour and find 50, which is the current to be used.

Fig. 51.—Table of Charging Rates.

rators, after having been in service, will not stand much handling and had better be thrown away. If it is thought worth while to keep them they must be immersed in water or weak electrolyte, and in reassembling the electrolyte must be put into the cells immediately, as wet wood separators must not stand exposed to the air for any unnecessary moment, especially when in contact with plates. Storage batteries always should be stored in a dry place, preferably in one where the temperature will never fall below 40° Fahr. Storage-battery solution or electrolyte varies greatly in density between the points of complete charge and complete discharge. When completely discharged the electrolyte of the average battery has a specific gravity of 1.14, and a sulphuric acid solution of 1.14 specific gravity has a freezing-point of about 10° Fahr. Therefore, if a completely discharged battery is allowed to stand where it is exposed to extremely low temperature it is quite possible for the electrolyte to freeze and the cells to be injured in consequence. However, as already pointed out, a battery for other reasons must not be allowed to stand in the discharged condition for any length of time. With increasing charge the density of the electrolyte increases until, when the charge is complete, it attains 1.28 specific gravity. The freezing temperature of the solution drops very quickly as the specific gravity increases, somewhat as follows:

Specific Gravity.	Freezing-Point Degrees.
1.14	+10
1.16	+ 5
1.175	− 4
1.20	−16
1.225	−36
1.25	−60
1.28	−85

Consequently, there is no possibility of a storage battery being injured by freezing in this latitude if it is kept in a fair state of charge. The freezing-points of electrolyte of different specific gravities are also shown in graphic chart at Fig. 53.

Charging Vehicle Batteries of Lead-Plate Type.—The following extracts on modern electric vehicle batteries are reproduced from an article by J. H. Tracy and with the permission of The

Electric Storage Battery Company, who issue the complete discussion in their Bulletin No. 156. These instructions do not apply to all types of lead-plate batteries, however, but refer to the latest vehicle types made by this company.

To the user of lead acid storage batteries in self-propelled vehicles, the steady improvement of recent years is hardly noticeable to the eye, although there has been an increase in the watt-hour capacity of the battery per unit of space and weight,

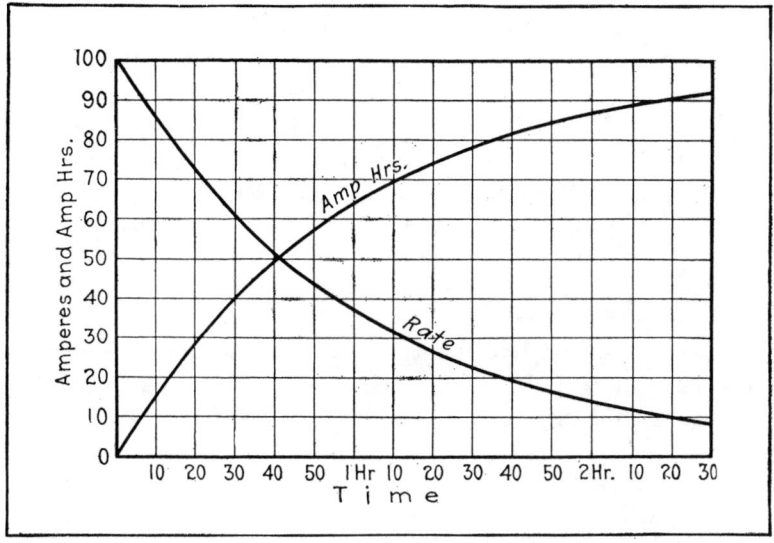

Fig. 52.—Theoretical Variation in Charging Rate When the Rate in Amperes Equals the Ampere-Hours Out of the Battery.

and also in the serviceable life of the battery. There has also been a marked advance in the permissible rates of charge and discharge, which has added so much to the flexibility of operation as to permit the use of much smaller batteries than would have been considered a few years ago in the same service. The following article explains the permissible rates of discharge and the behavior of batteries under operating conditions which a few years ago were considered prohibitive:

It can be safely said that high rates of discharge are in no way detrimental to modern lead battery plates. Batteries of the vehicle type are in regular operation under conditions in which practically all their work is done at rates which would empty the battery in ten minutes, and the same batteries would be sold to operate at the three-minute rate if there were a commercial demand for such operation. For such high rates of discharge, extra heavy terminals are provided to carry the current, no other changes being required.

It is well known that when discharged *continuously* at a constant rate the available ampere-hour capacity of a battery is a function of the rate of discharge, the available capacity being lower at the higher rates. This reduction in available capacity at the higher rates of discharge is due to depletion of the acid in the pores of the plates. The rate of this depletion is the difference between the rate of absorption by the plates of the acid that is in the pores of the plates and the rate at which this acid is renewed by diffusion with the other acid in the cell. It is the limit of this available acid that limits the capacity of the battery at high rates of discharge, and not any limitation in the plates themselves. It is, therefore, impossible to damage the plates by overdischarge at high discharge rates. In fact, very low rates of discharge should receive more careful consideration than very high rates.

In general, a battery may be charged at any time when a charge will be useful and at any rate which will not cause the temperature of the battery to exceed 110° F. and which will not cause the cells to gas freely except at low rates of charge. If these conditions can be watched no further directions or limitations need to be considered. As it is not always possible to watch these conditions, several methods of charging have been worked out which reduce the amount of attendance required while charging, and which permit the selection of the most economical way to charge the battery under any particular set of local conditions, while assuring that the above limitations will not be exceeded.

A general rule for determining the maximum permissible rate of charging a battery is: The charging rate in amperes must never exceed the ampere-hours out of the battery. Any method of

CHARGING VEHICLE BATTERIES

charging that keeps the charging current within this limit will not overheat the battery or cause it to gas. In applying this rule it is not necessary to reduce the charging rate below the "finishing" rate recommended by the battery manufacturers. If an ampere-hour meter is used on the vehicle, so arranged as to indicate the ampere-hours *out* of the battery, it also indicates at all times the maximum permissible charging rate. It will be noted that the maximum charging rates are no longer a function of the

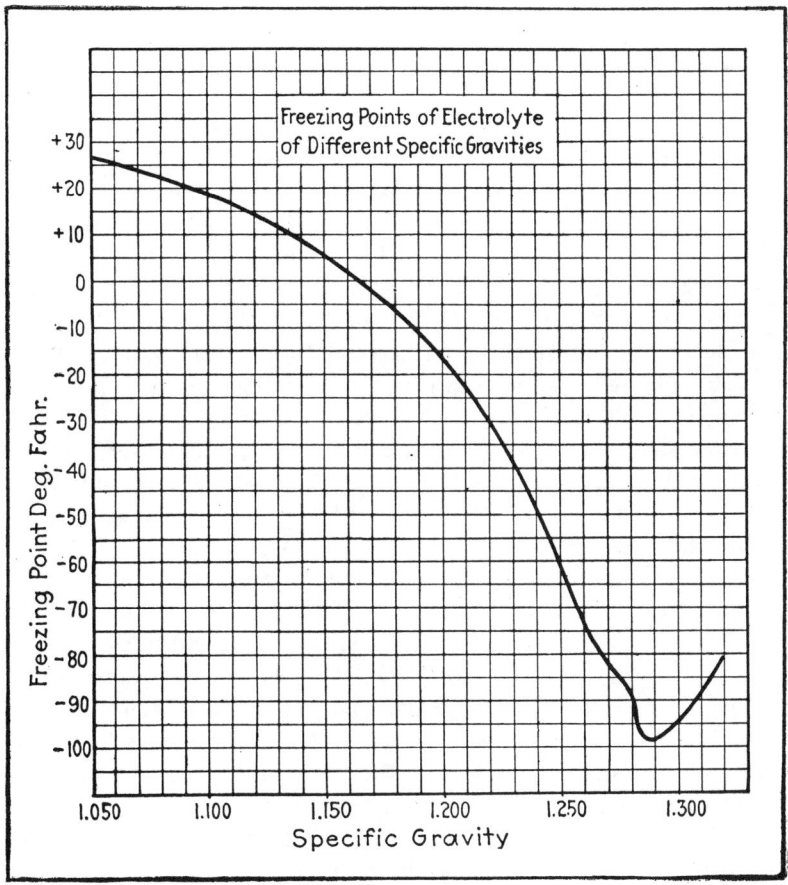

Fig. 53.—Freezing-Points of Battery Electrolyte.

size of the battery or its relative state of discharge, but depend only on the actual state of discharge. The curve in Fig. 52 shows in percentage the theoretical variation in charging rate and also in state of charge if a battery were charged strictly in accordance with this rule, and this represents the method by which a battery may be safely charged in a minimum time in regular operation.

It is evident that very wide latitude for proper charging is offered from which to select the best way to charge a battery under any given local conditions. If the vehicle is equipped with an ampere-hour meter the readings of this meter may be taken as the basis for selecting a charging rate which may be used for a particular length of time, so that at the end of that time the charging rate will be at the maximum permissible rate, at which time the rate should, of course, be reduced. It follows from the general rule for charging that, if $R =$ permissible charging rate, $D =$ ampere-hours out of the battery at the start of the charge (reading of the ampere-hour meter) and $T =$ time in hours until the current can be adjusted, then $R = D \div (+ T) =$ maximum permissible charging rate for T hours, and at the end of T hours the charging rate will equal the reading of the ampere-hour meter. This value can again be divided by $1 + T$ for the new charging rate, and so on until the charge can be finished at the finishing rate. If it is desired to charge the battery rapidly, the time T should be taken as short as possible. For convenience Table I, given in Fig. 51, calculated from this formula, is given. This table is of use not only in the charging-room, but also for the determination of the best manner for charging vehicles under any contemplated conditions and for the selection of charging equipment to meet the requirements of these conditions.

Ampere-hour Meter Indications as Basis for Charging: It should be carefully noted that if an ampere-hour meter is made the basis for charging a battery, care must be taken to be sure that the meter indicates as nearly as possible the real state of charge of the battery. An accurate record of the ampere-hours discharge from a battery does not give an accurate measure of the ampere-hours necessary to fully recharge it, for there are certain variable losses in the battery which the ampere-hour meter cannot measure. In fact, there is no accurate way to predetermine exactly how many ampere-hours charge may be necessary to fully charge a battery, nor is it necessary in ordinary service that the battery be really completely recharged daily. An ordinary clock is not an accurate instrument for measuring time, yet if it is set correctly occasionally it is sufficiently accurate for ordinary purposes. It is the same with an ampere-hour meter. It is necessary that a battery be fully charged occasionally, say, once a week, if the battery is subjected to hard daily use, as on a commercial truck, and this furnishes an opportunity to set the ampere-hour meter.

A battery is fully charged only when all the sulphate has been driven out of the plates into the electrolyte, and this is most easily indicated by the specific gravity of the electrolyte. As long as sulphate is being thrown out of the plates into the electrolyte during charge, the specific gravity of the latter must continue to rise, and when the rise stops the battery is fully charged. Most battery manufacturers recommend that a battery be given such a charge (called an equalizing charge), regardless of the indication of the ampere-hour meter, once a week or once in two weeks. When it is known that the battery is full, the charge is discontinued and then the meter is set to indicate a full battery, and the meter is then a sufficiently accurate indicator of the state of battery charge to be used for a week or two weeks until another equalizing charge is given the battery, when the meter should again be set.

Ampere-hour meters require cleaning and regulation at intervals as does a clock, and if they are treated in this manner they are of great assistance in the proper handling of a battery. These meters are frequently furnished with a contact-making device, so arranged as to interrupt the charging circuit when the meter indicates that the battery is fully charged, and this is a valuable protection again unnecessary charging and gassing of the battery during ordinary operation. This tripping device should, of course, be disconnected during the equalizing charge.

CHAPTER V

Uses of Storage Batteries—Internal Combustion Engine Ignition—Automobile Starting and Lighting Systems—Shifting Gears by Electricity—Electric Pleasure and Commercial Automobiles—Isolated Lighting Plants — Train Lighting — Storage-Battery Locomotives — Battery-Driven Street Cars—Submarine Boats—Miscellaneous Marine Applications—Railway Switch and Signal Service—Stand-by and Booster Service—Drawbridge Operation—Mine Lamp Battery.

Uses of Storage Batteries.—In this chapter the writer intends to describe briefly the most interesting of the many possible applications of storage batteries, some of which are unusual, to say the least. A rather flexible imagination is needed to consider that the cheerfully lighted farm home; the lifting of a massive drawbridge; the roaring start of a powerful automobile or hydro-aeroplane engine and the noiseless movement of the electric automobile are all accomplished by the same agency. Storage batteries propel numerous electric launches on the water's surface, and are the sole power for the submarine lurking in its depths. Hundreds of palatial yachts are illuminated by this means, and the penetrating shaft of light from the miner's lamp is produced by the same energizing source. A wireless call for help from a ship in distress, with its life-saving possibilities, is produced from a few cells of a storage battery. The headlights of the myriad automobiles now using our highways after nightfall attest to the practical value of this current-producer.

The magnitude of the industry and the great possibilities of the field for storage batteries can be only briefly touched upon, but it is evident, from a perusal of the following list, that they can be used anywhere a dependable source of current is required. We find storage batteries used in: Central lighting and power stations; electric railway power houses; municipal and office building lighting and power; steel mills; electric hoist and elevators;

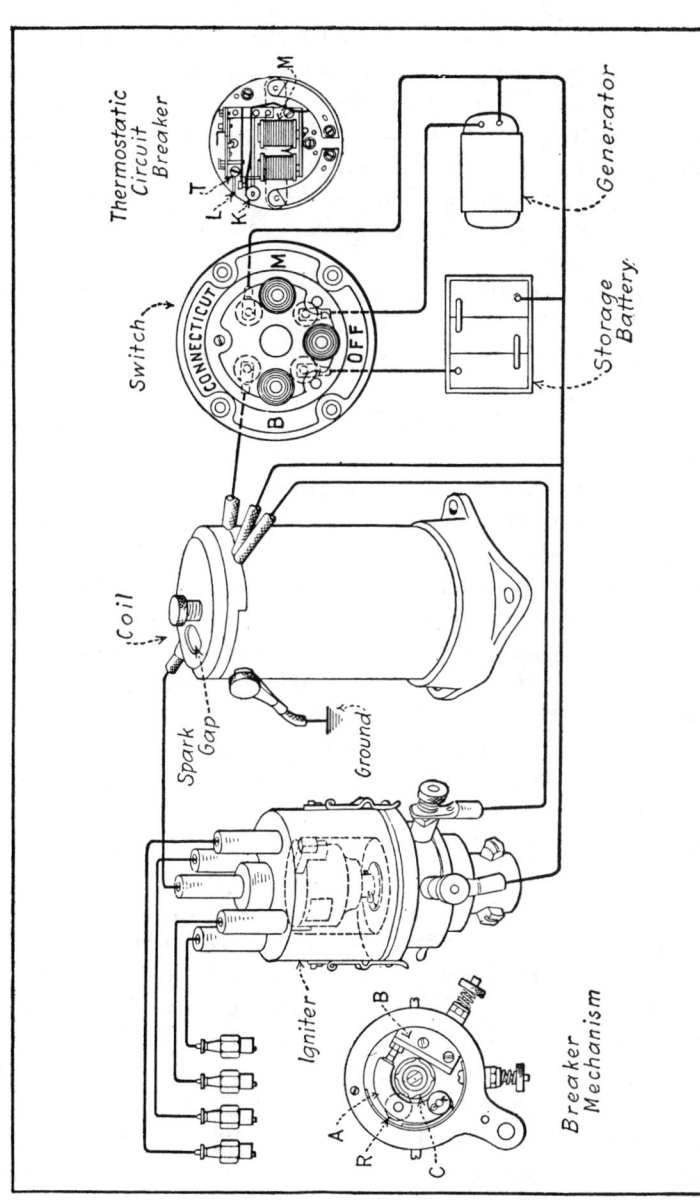

Fig. 54.—Connecticut Closed-Circuit Ignition System Uses Storage-Battery Current.

isolated lighting plants for hotels; suburban residences and farms; railway switch and signal service; railway car lighting; interlocking switch service; United States Government submarine and gun-firing service; telephone, telegraph, wireless and fire-alarm service; laboratory and school work; electroplating; automobile engine-starting; gas-engine ignition; automobile lighting; electric trucks and pleasure cars; street railway cars; electric launches and mine and industrial locomotives.

The Storage Battery for Gasoline-Engine Ignition.—Because of the almost universal employment of electricity for lighting and starting ystems, the battery ignition system has been improved materially, inasmuch as the storage battery supplying the current is constantly charged by a generator. A number of systems have been devised, these operating on two different principles, the open circuit and the closed circuit. An example of the closed-circuit system is shown at Fig. 54, and is of Connecticut design, the complete ignition system consisting of a combined timer and high-tension distributor, a separate induction coil and a switch. The system is distinctive in that the timer is so constructed that the primary circuit of the coil is permitted to become thoroughly saturated with electricity before the points separate, with a result that a spark of maximum intensity is produced. The action is very much the same as that of a magneto on account of the saturation of the winding. Another feature is the incorporation with the switch of a thermostatically operated electro-magnetic device, which automatically breaks the connection between the battery and the coil should the switch be left on with the motor idle.

The contact-breaker mechanism consists of an arm A carrying one contact, a stationary block B carrying the other contact, a fiber roller R, which is carried by the arm A, and operated by points on the cam C, which is mounted on the driving-shaft. Normally, the contacts are held together under the action of a light spring. As the four cams, which in touching the roller R raise the arm and separate the contacts, are 90 degrees for a four-cylinder motor, the period of saturation of the coil or the length of time the current flows through it to the battery is sufficiently long so that when the points have separated, the current, which

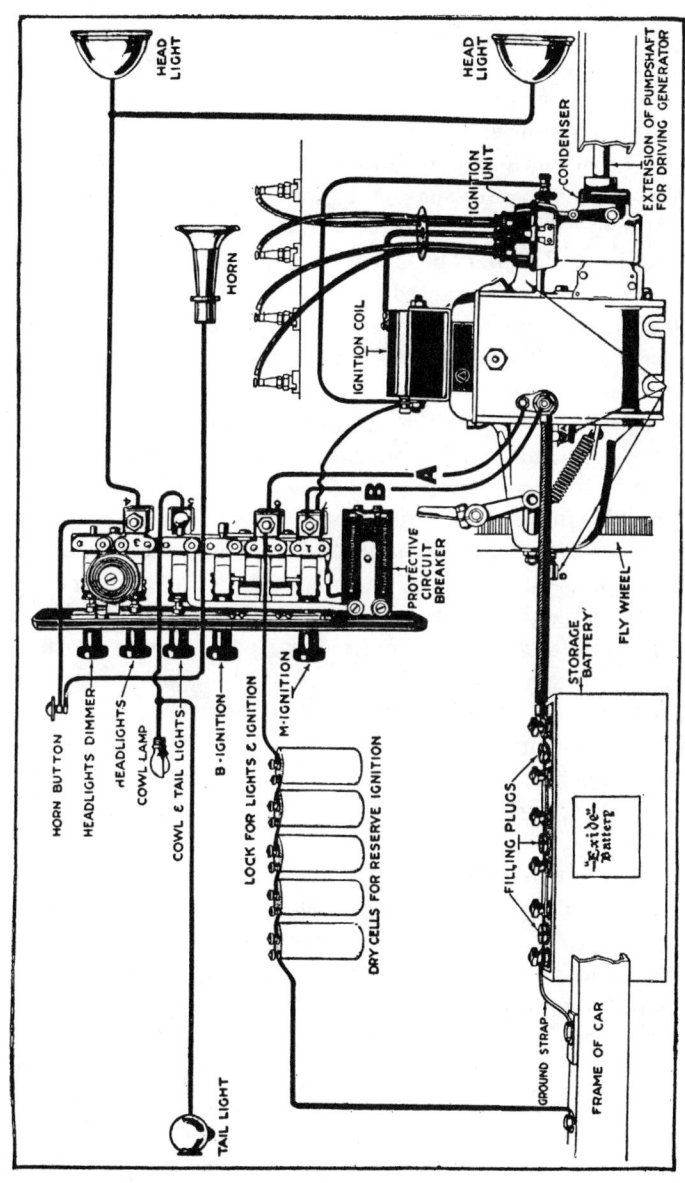

Fig. 55.—The Delco Ignition, Motor-Starting and Car-Lighting System.

has "piled" up, induces an intensely hot spark at the plugs. This is an advantage, inasmuch as it insures prompt starting and regular ignition at low engine speed, as well as providing positive ignition at high engine speed.

The thermostatic circuit-breaking mechanism is very simple. This consists of the thermostat T, which heats when the current passes through it for from thirty seconds to four minutes without interruption, and thus is bent downward, making contact with the contact L. This completes an electrical circuit, which energizes the magnets M, causing the arm K to operate like the clapper in an electric bell. This arm strikes against the plate, which releases whichever of the two buttons in the switch may be depressed.

As will be observed, the transformer coil provided has five terminals. One of these is connected directly with the ground, the other leads to the central secondary distributing brush of the timer-distributor. Of the three primary leads, one goes to the switch, one to the wire leading from the storage battery to the timer, and one directly to a terminal on the timer. The switch is provided with three buttons, the one marked B being depressed to start the engine, as the ignition current is then drawn from the storage battery. After the engine has been started the button marked M is pressed in, this taking the current directly from the generator. To interrupt ignition the button "off" is pressed in, this releasing whichever of the buttons, B or M, is depressed. Four wires run from the distributor section of the igniter to the spark plug.

One of the most popular of the combined starting, lighting and ignition systems is the Delco, which is shown at Fig. 55. For the present we will concern ourselves merely with discussing the ignition functions of the system, leaving the self-starting and electric-lighting features for consideration later. Current is produced by a one-unit type motor-generator, although the windings of the device when operated as a motor or a generator are entirely separate. The ignition current is obtained either from a storage battery, which is kept in a state of charge by the generator, or from a set of dry cells which are carried for reserve ignition. The ignition system consists of a one-unit non-vibrator coil,

STORAGE BATTERY FOR IGNITION

sometimes attached to the top of the motor generator, though it may be placed at any convenient part of the car, and a dual automatic distributor and timer usually included as a part of the device as shown. When ignition current is supplied from the lighting circuit the current passes from the storage battery through a switch and out to the low-tension winding of the coil, from whence it passes to the timer and from there to the frame, where it is grounded. The high-tension current generated in the coil runs to the distributor, where it is switched to the spark plug in the different cylinders in turn.

The essential elements of any electrical ignition system, either high or low tension, are: First, a simple and practical method of current production; second, suitable timing apparatus, to cause the spark to occur at the right point in the cycle of engine action; third, suitable wiring and other apparatus to convey the current produced by the generator to the sparking member in the cylinder. The important part the storage battery plays in the gasoline automobile can be readily understood by the reader.

Storage Battery for Starting Automobile Motors.—One of the most recent applications of the storage battery is in starting gasoline engines used in automobiles. The storage battery has made the old hand crank obsolete, and has provided a convenient lighting system as well as a positive motor-starting means. The parts of a two-unit starting and lighting system are shown at Fig. 56. This system is sometimes called a "three-unit" system, on account of having a source of independent current supply for ignition purposes. As will be observed, the generator is driven from the motor crankshaft by silent chain connections, one of the terminals passing through the cut-out device and to the storage battery, the other terminal running directly to the storage-battery terminal, having a short by-pass or shunt wire attached to the cut-out. All the time that the engine is running the generator is delivering electricity to the storage battery.

It will be observed that the storage battery is also coupled to the lighting circuits, which are shown in a group at the right of the illustration, and to the electric-starting motor as indicated. One of the storage battery terminals is joined directly to the switch

128 STORAGE BATTERIES SIMPLIFIED

terminal by a suitable conductor, the other goes to one of the terminals on the starting motor, while the remaining terminal of the starting motor goes to the switch. In this system, when the small sliding pinion is meshed with the flywheel gear, the switch is thrown on simultaneously, and the current that flows from the storage battery through the windings of the starting motor rotates the engine crankshaft by means of reduction gears shown. As soon as the engine starts the foot is released and a spring pulls

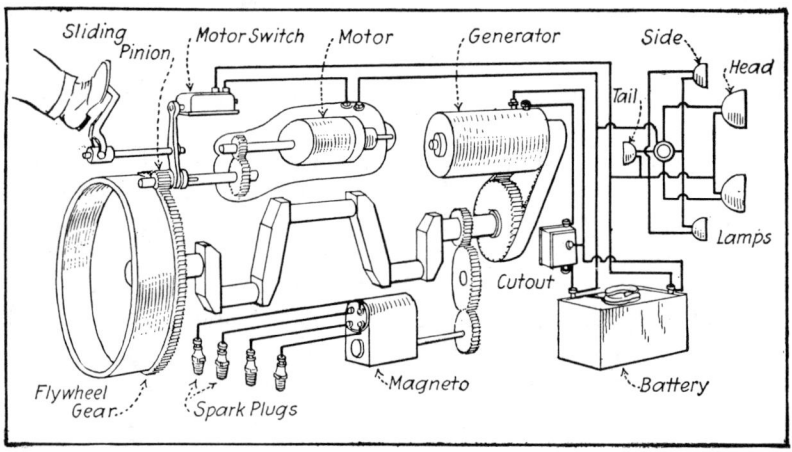

Fig. 56.—Diagram Showing Components of Two-Unit Starting and Lighting System.

the switch out of contact, and also disengages the sliding pinion from the flywheel gear. Electrical starting systems are usually operated on either six- or twelve-volt current, the former being generally favored because the six-volt lamps use heavier filaments than those of high voltage, and are not so likely to break, due to vibration. It is also easier to install a six-volt battery, as this is the standard voltage that has been used for many years for ignition and electric lighting purposes before the starting-motors were applied.

In referring to a system as a one-unit system of lighting, starting and ignition, one means that all of these functions are incorporated in one device, as in the Delco system at Fig. 55. If one

STORAGE BATTERY FOR STARTING MOTORS

unit is used for generating the lighting and starting current, and also is reversible to act as a motor, but a separate ignition means is provided, such as a high-tension magneto, the system is called a "two-unit" system. The same designation applies to a system when the current generating and ignition functions are performed by one appliance, and where a separate starting-motor is used. The three-unit system is that in which a magneto is employed for ignition, a generator for supplying the lighting and starting current, and a motor for turning over the engine crankshaft.

The generator, as is apparent from its name, is utilized for producing current. This is usually a miniature dynamo patterned largely after those that have received wide application for generating current for electric lighting of our homes and factories. The generators of the different systems vary in construction. Some have a permanent magnetic field, while others have an excited field. In the former case permanent horseshoe magnets are used, as in a magneto. In the other construction the field magnets, as well as the armature, are wound with coils of wire. In all cases the dynamo or generator should be mechanically driven from the engine crankshaft, either by means of a direct drive, by silent chain, or through the medium of the timing or magneto-operating gears. Belts are apt to slip and are not reliable.

All the current produced by the generator and not utilized by the various current-consuming units, such as the lamps, ignition system, electric horn, etc., is accumulated or stored in the storage battery, and kept in reserve for starting or lighting when the engine is not running or for lighting and ignition when the car is being run at such low speed that the generator is not supplying current. Storage batteries used in starting systems must be of special design in order to stand the high discharge and to perform efficiently under the severe vibration and operating conditions incidental to automobile service. The storage battery may be installed on the running-board of the automobile, under the body, or under the front or rear seat, the location depending upon the design of the car and the degree of accessibility desired, as shown at Fig. 57. The best practice is to set the storage battery in a substantial carrying case held by rigid braces attached to the frame side and cross members. If the battery should be set under the tonneau floor boards, a door must be provided in these to give ready access to the battery.

The starting-motor, which takes the place of the common hand-crank, is operated by current from the storage battery, and the

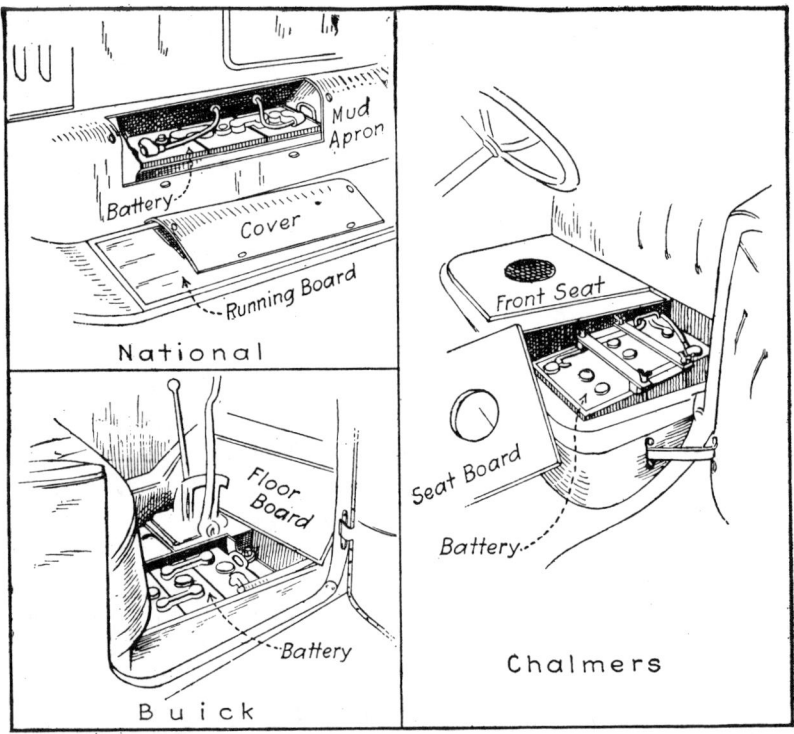

Fig. 57.—How Storage Batteries are Installed in an Automobile When Used for Starting and Lighting Current.

high-speed armature rotation is reduced to the proper cranking speed by reduction gears of the different forms, to be described in proper sequence. The construction of the starting-motor is practically the same as that of the dynamo, and it operates on the same principle, except that one instrument is a reversal of the other.

In order to secure automatic operation of a lighting and start-

STORAGE BATTERY FOR STARTING MOTORS 131

ing system several mechanical and electrical controls are needed, these including the circuit breaker, the governor, which may be either mechanical or electrical, and the operating switches. The circuit breaker is a device to retain current in the storage battery under such conditions that the battery current is stronger than that delivered from the generator. If no circuit breaker was provided the storage battery could discharge back through the generator winding. The circuit breaker is sometimes called a "cutout." The circuit breaker is usually operated by an electro-magnet, and may be located either on the generator itself or any other convenient place on the car, though in many cases the circuit breakers are usually mounted on the back of the dashboard. This device is absolutely automatic in action and requires but little attention.

The governors are intended to prevent an excessive output of current from the generator when the engine runs at extremely high speed. Two types are used: one mechanical, operated by centrifugal force, and the other electrical. The former is usually a friction-drive mechanism mounted on the generator shaft, which automatically limits the speed of the dynamo armature to a definite predetermined number of revolutions per minute. The maximum current output is thus held to the required amount independently of the speed at which the car is being driven. The use of this device minimizes the possibility of overheating the generator overcharging the battery at high car speeds. The electrical system of governing does not affect the speed of the armature, but controls the output of the generator by means of armature reaction and a reversed series field winding. The governors usually permit a maximum generator output of from ten to twelve amperes, though the normal charging current is less than this figure.

In practically all systems an amperemeter is mounted on the dash so that it can be readily inspected by the driver, this indicating at all times the amount of current being produced by the dynamo or drawn from the battery. If the indicating needle of the amperemeter points to the left of the zero point on the scale, it means that the battery is furnishing current to the lights or other current-consuming units or discharging. When the needle points to the other side of the scale, it means that the generator is delivering current to the battery which is charging it; the amount of charge or discharge at any time can be read from the scale on the face of the amperemeter. Some of these instruments have the words "charge" and "discharge" under the scale in order to enable the operator to read the instrument correctly.

Another important element is the lighting switch, which is

Fig. 58.—Method of Installing Starting and Lighting Battery in Automobile Frame.

usually mounted at some point within convenient reach of the car driver. This is often placed on an instrument board on the back of the cowl in connection with other registering instruments. As ordinarily constructed, the switches are made up of a number of units, and the wiring is such that the head, side and tail lamps may be controlled independently of each other. For simplicity and convenience of installation, the switch is usually arranged so that all circuits are wired to parallel connecting members, or "busbars," placed at the rear of the switch.

STORAGE BATTERY FOR STARTING MOTORS

APPROXIMATE CURRENT FOR LAMPS

Voltage of Lamps	Amperes per Candle Power in Tungsten Lamps	
	Head or Other Large Lamps	Side, Tail or Other Small Lamps
6	0.17	0.21
8	0.125	0.16
12	0.085	0.1

AMPERE-HOUR CAPACITY

Capacity at Normal Temperature of "Exide" Type LX and SX Cells when Discharged Continuously at Rates Required for Ordinary Lighting and Ignition Service.

	Size of Cell					
	LX-5	LX-9	LX-13	LX-17	SX-9	SX-13
Ampere Hours in Ignition Service..................	50	100	150	200	80	120
Ampere Hours when Discharged intermittently for Lighting*................	40	80	120	160	64	96
Ampere Hours at 1 Ampere Rate....................	39	86½	136	186	68	109
Ampere Hours at 1½ Ampere Rate....................	36½	81½	130	180	64	102
Ampere Hours at 3 Ampere Rate....................	32	73	118	164	57	91
Ampere Hours at 5 Ampere Rate....................	28	66	108	151	51	84
Ampere Hours at 7½ Ampere Rate....................	25	60½	100	141	46	77
Ampere Hours at 10 Ampere Rate....................	23	56	94	133	43	73

* This capacity may be obtained by intermittent discharges of a few hours each during a period of three days or more.

How Storage Battery Shifts Gears.—A new system of gear shifting has been recently developed which depends on the use of electric current to shift the gears instead of the usual hand lever. The steering wheel is shown at Fig. 59, with the various speed-changing buttons let into a box attached to the steering post. The wiring is outlined. The operation of shifting a gear is very simple, consisting merely of depressing the clutch pedal and pressing down on the switch button marked with the gear ratio desired. The system is not complicated, the gears being controlled by solenoid coils, one being used for each forward speed and one for reverse. Two switches are utilized between the battery and the coils, a knife switch controlled by the clutch pedal and a push-button located on the steering wheel. All changes of gears are controlled by the knife switch, and the push-buttons on steering wheel merely arranges the circuit for the particular speed

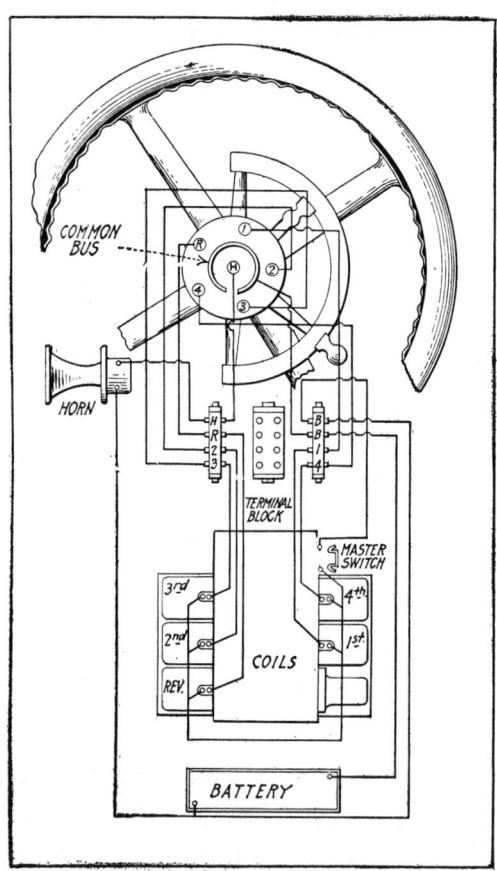

Fig. 59.—Wiring Diagram Showing the Method of Connecting the Vulcan Electric Gear Shift with the Battery and Control Switch.

HOW STORAGE BATTERY SHIFTS GEARS

desired. The first movement is the regular operation of the clutch, but a continued operation of the clutch lever actuates the knife switch.

Current flows from the battery through the solenoid coil and pulls a plunger against a magnet with a force which is given as

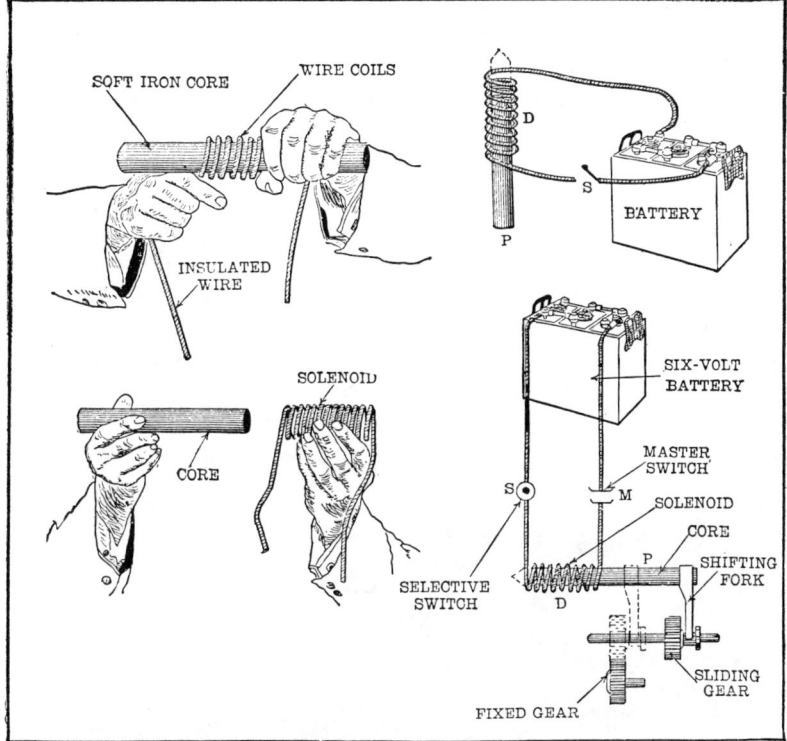

Fig. 60.—Simplified Diagrams Showing How Current Passed Through the Solenoid Will Draw in Iron Core Piece, Which May be Made to Shift the Gears.

40 to 100 pounds. This energy is transmitted through an arm to the gear-shifting fork and gear in exactly the same manner as if the gears were operated with a hand lever. The plungers are normally in a neutral position. When the button is pressed on the control member, current passes through the coil around one of the plungers, drawing it against the magnet. It is said that

136 STORAGE BATTERIES SIMPLIFIED

the current required to make the shift is about 17 amperes, and it is claimed, further, that three hundred speed changes may be made with less current consumption than is required in starting the motor with an electric-starting device. An advantage claimed for this electric gearshift is that the gears cannot be stripped, for the reason that the clutch must be disengaged before a shift can be made, and the gears are always in neutral before the coils can accomplish the change.

Batteries for Electric Automobiles.—Any practical form of storage battery may be used for automobile propulsion. Either the Edison or lead-plate type batteries prove satisfactory for this work. The lead-plate forms have thin plates to keep the weight down and make for quicker charging and discharging. The usual discharge rate is about two volts per cell, the amperage depending upon the resistance to vehicle motion. It is said that the annual maintenance cost of the ordinary lead-plate cell is about 60 per cent. of its original cost. A special vehicle type, called the "Iron Clad" (previously described), will cost about 30 per cent. of its initial purchase price annually to maintain it in proper condition. The Edison battery occupies about one-third more space

Fig. 61.—Electrically Propelled Taxicab, Having Part of Storage Battery Under Hood and Part Under Rear Seat.

BATTERIES FOR ELECTRIC AUTOMOBILES 137

than a lead-plate battery of equal capacity, and costs about two and one-half times as much, but it is said to retain its full capacity for three or four years. The lead-plate type will deliver about 10 watt-hours per pound weight; an Edison battery will give nearly 14 watt-hours per pound.

The usual number of lead-plate cells provided to charge from a 110-volt circuit is forty-two or forty-four, having 11, 13 or 15 high-capacity plates per cell, while sixty-cell Edison batteries are

Fig. 62.—Industrial Truck, Using Battery Power at A. High-Speed Electric Roadster Automobile at B.

supplied to meet the same conditions. The battery weight of the average electric vehicle is about 35 per cent. of the total unloaded weight. In order to insure a rate of discharge that will not injure the batteries, electric vehicles are usually geared for moderate speeds, seldom more than 18 to 20 miles per hour for pleasure or passenger-carrying cars on pneumatic tires. Heavy trucks seldom run faster than 6 miles; medium-capacity commercial vehicles may run 10 or 12 miles per hour. A number of electric vehicle-makers have adopted a 24-cell battery, using motors wound for about 48 volts.

The motors used are almost always of the series-wound type, because they provide more power for starting. The usual pleasure-car size is a motor of 2 k.w. rating, or about $2\frac{3}{4}$ horsepower. Vehicle motors have an overload capacity of 200 to 300 per cent., though, of course, this is practical for only relatively short periods of operation. The speeds are controlled by different combinations of the motor *field* windings.

In some electric vehicles the various car speeds were obtained by parallel and series parallel combinations of the battery cells. From four to six forward speeds are generally provided, a simple reversing arrangement making it possible to have the same number of reverse speeds if necessary, though this is not always done. Another scheme of control is to have the motor field windings or coils so wired that they may be put in series or in parallel groups. A resistance is used in securing the first speed with the field coils in series. The second speed is obtained by cutting out the resistance and leaving the fields in series. Third speed is obtained by shunting the resistance across the fields, which are still in series. The fourth speed is obtained by leaving the field coils in parallel connection and with resistance out, the fifth speed by shunting the resistance in with the parallel connected fields. The parallel connection gives greater speed, the series field connection more pulling power.

Electric motors for electric trucks usually have about one kilowatt capacity per ton load added, the minimum being 2 k.w. on a one-ton truck; thus a two-ton truck will need a 3 k.w. motor, a four-ton truck 4 k.w., and 5 k.w. for a five-ton vehicle. The usual

Fig. 63.—Industrial Truck for Factory Work Propelled by Storage Battery.

range on good roads is about 40 miles for a truck and 75 to 100 miles for pleasure cars on one battery charge. Vehicle batteries are generally carried in hard rubber jars and are nearly always of the sealed type on account of the liability of splashing the electrolyte when the car is operated over ordinary highways. The cells are grouped in trays for easy handling, and all connections and couplings, terminals and battery straps are of unusually rugged design.

The present tendency in lead-battery design seems to be toward the use of more thin plates, as fifteen are furnished as a standard equipment more often than a smaller number. Such an equipment will give 180 ampere-hours, and in some types may develop fully 200 ampere-hours. The reasons for the increasing adoption of the thin-plate lead battery are, first, an augmenting demand for more speed and greater range of action per charge, and secondly, a realization on the battery-maker's part that the life of a thin plate is equal to that of the heavier ones if the installation is properly made. One standard jar size now being produced will work with from 11 to 15 plates. The high-ribbed type of jar is used, having at least three inches below the bottom of the plates for sediment space, this reducing the amount of washing and necessary cleaning out of the cells. The jars are assembled side to side in trays, with the plate surfaces set at right angles to the direction of car movement. One row of cells is mounted in each tray, these being set lengthwise in the battery compartment. This arrangement is said to reduce jar breakage.

Isolated Lighting Plants.—Many makers of storage batteries have developed types for use in residence and factory lighting where central station power is not available. The advantages of electricity for the supply of light, heat and power have led to a demand for the satisfactory and economical operation of isolated plants. The marvelous development of the internal combustion gas, gasoline or kerosene engine and the improvements in the various forms of lamps have created an active interest in the application of the storage battery to isolated plants of moderate size, it now being recognized where uninterrupted twenty-four-hour service is desired a storage battery is an absolute necessity. Usu-

Fig. 64.—Electrically Propelled Truck, With Battery Compartment Under Driver's Seat.

Fig. 65.—Plan and Side Views of Electric Truck Chassis, With Battery Box Removed, Showing Simplicity of Mechanism.

ISOLATED LIGHTING PLANTS 143

ally in isolated plants the engine and generator capacity is sufficient for the total number of lamps connected, although they are all seldom in use at one time, consequently the plant operates at but partial load during the total lighting hours. This means low efficiency, poor regulation and high fuel costs. The installation of a storage battery corrects this weakness by permitting the operation of the generator at the full or the most economical load for

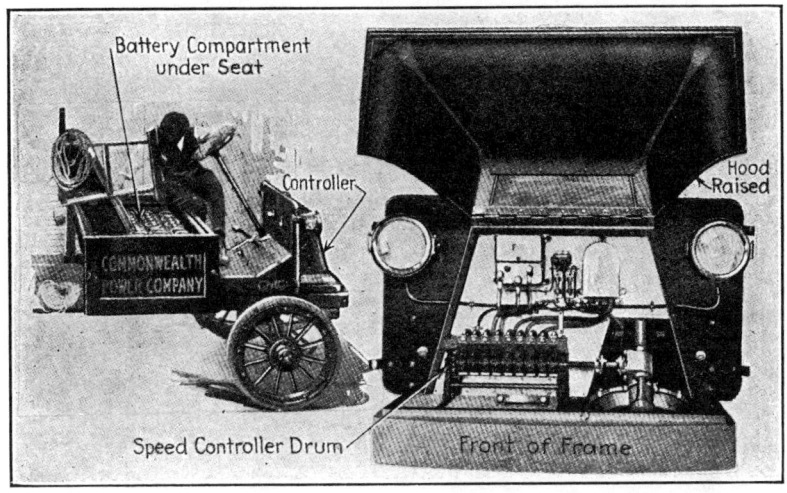

Fig. 66.—View at Left Shows Battery Location; at Right the Controller Compartment is Opened to Show Accessibility of Wiring.

a few hours and then shutting down entirely, the battery providing the current for the balance of the time.

In many cases it can be so arranged that the generator need be operated only every second or third day, and then at the most convenient time. No additional labor is required; in fact, this cost is usually lessened, while the fuel cost and maintenance and repair expense are much reduced. By taking current directly from the battery, steady lights are obtained and the noises of engine at night avoided. When on special occasions unusual lighting is required, the battery may be discharged in parallel with generator, and demands equal to the combined capacity of battery and

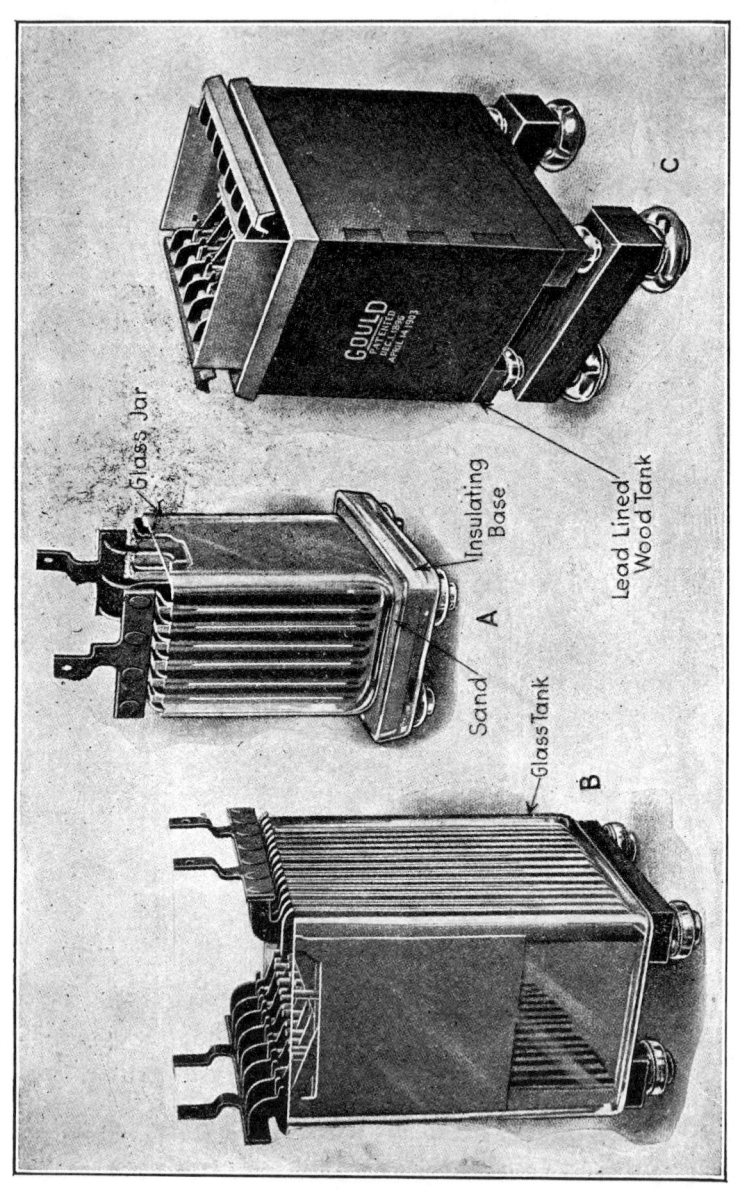

Fig. 67.—Types of Gould Storage Batteries for Isolated Lighting Plants.

generator may be supplied. The generating equipment may be stopped at any time for adjustment or repair without interrupting the service, the battery being available for unexpected demands for power. The great advantage of electric current for operating pumps and other machinery at a distance from the engine and generator cannot be overestimated. The employment of electricity for driving fans, heating curling-irons, cooking, etc., is also of great convenience, and by installing batteries current is available at all times.

The following rules for battery selection and methods of installation and operation are reproduced from a bulletin issued by the Gould Storage Battery Company, describing batteries for isolated lighting plant use.

Selection of Battery.—The number of cells is determined by the voltage of the system and is entirely independent of the size of the individual cells.

Isolated plants of the various voltages require batteries of the number of cells given in the following table:

Voltage of System	Number of Cells	Voltage of System	Number of Cells
110	60	220	120
115	64	230	126
125	70	250	138

The size of the individual cells is determined by the number of lamps, their candle-power and efficiency, and the length of time they must be supplied on one discharge. For ordinary purposes it is sufficiently accurate to estimate the energy taken by a 16-candle-power carbon filament lamp as 55 watts (110-volt, 16-candle-power lamp taking one-half ampere) and lamps of other candle-power on a proportionate basis. By using tungsten filament or nitrogen-filled bulbs, the current consumption may be materially reduced for a given candle-power, and outfits of lower voltage will give satisfactory light. For example, a recently developed, small-capacity lighting outfit for farm use uses but 16 cells of battery by operating 30-volt tungsten filament lamps. To simplify figuring, we will consider a voltage that is in common use in places where a central station furnishes power or 110 volts.

Storage batteries are rated in "ampere-hours," which defines their capacity and is the product of the number of amperes discharge and the number of hours such discharge can continue. The capacity at the eight-hour rate is considered the normal. As the ampere discharge is increased above the normal or eight-hour rate, the ampere-hour capacity decreases, as will be seen by the following example:

Rate, Hours	Ampere Discharge	Capacity Ampere Hours
8	12½	100
5	17½	87½
3	25	75
1	50	50

Thus, while 12½ amperes may be obtained for eight hours, or 100 ampere-hours, if the discharge be made at 25 amperes it can be continued for but three hours, or 75 ampere-hours; the remaining capacity of the battery is, however, available at a lower rate. On discharge at less than the eight-hour rate, the capacity of the battery is slightly greater, but the increase is small, and for ordinary calculation it is best to consider the capacity at rates lower than the eight-hour, the same as the eight-hour capacity.

The size of a 110-volt battery can be approximately determined by the method outlined in the following example, the conditions being that the battery will be charged at any time during the day convenient to operate the generator, and that the battery will be able to furnish current for lamps according to the following schedule:

Time	Number of Lamps		Amperes	Number of Hours	Amperes Hours
5 P.M. to 10 P.M.	Twenty	16 c. p.	10	5	50
10 P.M. to 6 A.M.	Two	8 c. p.	½	8	4
6 A.M. to 8 A.M.	Six	16 c. p.	3	2	6
					60

The last discharge rate is three amperes, and there will be required a battery of sufficient size to furnish 60 ampere-hours at a three-ampere rate. This being less than the eight-hour rate, we require a battery having a normal rating of 60 ampere-hours. By referring to the tables given in the Gould catalog it will be seen that Type M-307 will give 7.5 amperes for eight hours, or 60 ampere-hours at normal rating. Battery required, 60 cells, Gould Type M-307.

The above example shows a condition where the full normal capacity of the battery is used in carrying the load. Under some conditions this is not possible, i. e., where the latter part of the discharge is at a high rate; and it is advised that the battery company check the size of battery before the final decision.

Methods of Operation.—The principal function of a storage battery in small plants being the furnishing of current for a considerable period of time, such as the night load of a residential plant, the operation of the battery consists of cycles of charge and discharge covering practically the capacity of the battery. The problem of operation, therefore, resolves itself into two parts:

First—*Voltage control during discharge:* Under ordinary operating conditions, it is desirable to maintain practically constant voltage on the lighting circuits—hence, as the voltage of a storage battery varies during discharge, various methods have been developed to compensate for the changes in the battery voltage. A fully charged battery which, standing idle, shows about 2.1 volts per cell, will show while discharging at the eight-hour rate about 1.8 volts at the latter part of the discharge, and somewhat less at higher discharge rates. For isolated plants one of the following methods is usually employed: (A) Resistance control. (B) End-cell control.

Second—*Control of charging current:* To fully charge a battery it is necessary to raise the voltage, as the charge proceeds, from about 2.2 volts at start to approximately 2.62 volts per cell at the completion of the charge. This is usually accomplished in one of the following ways: (A) Normal voltage charge. Resistance control. (B) High-voltage charge directly from generator. (C) Shunt-booster charge.

148   STORAGE BATTERIES SIMPLIFIED

As the selection of a proper lighting battery should not be undertaken without consulting the battery maker, any reader wishing more involved, technical explanation of any of the methods described can obtain same by consulting the engineers of whatever battery maker he may select. Typical lighting batteries of Gould manufacture are shown in Fig. 67, which outlines the construction of the glass jar, glass tank and wood-tank types.

Details of Installation.—Gould storage batteries for light and

Fig. 68.—Typical Isolated Lighting Battery for Medium-Capacity Plant.

power plants are usually installed either in glass jars, glass tanks or lead-lined wooden tanks. The smaller type of cells are installed in glass jars resting on a bed of sand contained in a glass or wooden sand tray. The sand tray is supported by four glass insulators placed under the corners of the tray. Cells of medium capacity are usually installed in tanks of pressed glass, no sand trays being used, the glass tank resting on the glass insulators and separated therefrom by a small cushion of either lead or rubber to keep the hard surfaces out of contact. Cells of the glass

DETAILS OF INSTALLATION

jar (or glass tank) type are the easiest to install, as the plates are grouped at the factory. The plates of each cell are burned to common cross-bars and terminal straps, the negative plate to one cross-bar and the positive to the other, forming respectively the negative and positive "groups." The two groups with separators form what is known as the "element." The cells of the glass jar types are joined by bolting the lead terminal straps together by means of lead-covered brass bolts.

Lead-lined wooden tanks are often used for plants of medium size and always for those of large size. Plates of this type of cell are grouped at the place of installation by "lead-burning" the positive plates of one cell and the negative plates of the adjoining cell to a common busbar. All types of cells are usually installed with the supporting insulators resting on wooden stringers, these stringers having been previously painted with two or three coats of acid-proof paint. Cells of the smaller types which are not too heavy are generally installed on two-tier wooden racks in order to save floor space. The larger cells are installed in one tier, the wooden stringers being supported by vitrified brick set upon the floor or by another set of glass insulators resting on vitrified tiles. A typical installation of glass jar cells joined to form a medium-capacity battery is illustrated at Fig. 68. In this, the battery is in a special room prepared for it, and the cells rest on wooden stringers, as previously described.

Switchboard and Fittings.—It will be evident that the switchboard is an important detail of the storage battery lighting plant. The Type E switchboard used with some of the Electric Storage Battery Company outfits is shown at Fig. 69, the view at the right showing the face of the switchboard, while that at the left shows the method of installing the switchboard in connection with the battery supporting platform. The electric-current generator can also be fastened to the platform and belt-driven from the internal-combustion engine ordinarily used as power. In one corner, the ampere-hour meter dial is shown as an inset. Its use will be more fully described in proper sequence. The switchboard illustrated is utilized in connection with low-voltage house-lighting plants, and is simple, yet complete. The general arrangement has been

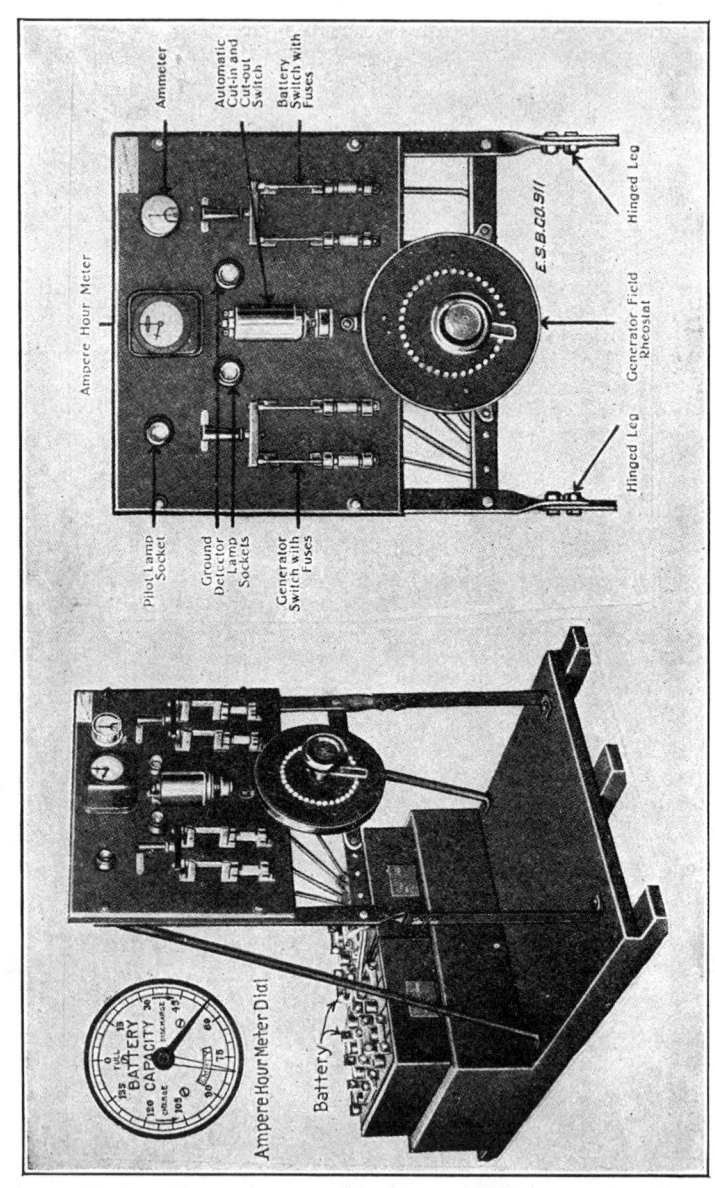

Fig. 69.—Typical Isolated Lighting-Plant Switchboard, Showing Fittings Needed.

SWITCHBOARD AND FITTINGS

worked out with great care by competent engineers who are thoroughly conversant with small-plant switchboard requirements as the result of experience gained in the installation and operation of great numbers of such plants.

The arrangement permits:

1. Lights to be run from battery only, or
2. Lights to be run from generator while generator is also charging the battery, or
3. Lights to be run from generator and battery in parallel, the battery assisting the generator, or
4. Lights to be run from generator only.

In the ordinary operation of the plant, no hand manipulation of switches or any part of the switchboard apparatus is at any time necessary except adjustment of the generator current by means of the generator field rheostat. Hand manipulation of switches is superseded by the automatic operations of the automatic cut-in and cut-out switch. This switch is simple, durable and reliable. It has no adjustments, because it needs none. Made in two sizes: 30-ampere, 32-42 volts, and 60-ampere, 32-42 volts. Panel of black oil-finished slate, 1 x $23\frac{1}{4}$ x 18 inches, on heavy strap-iron frame. Generator and battery switches are double-pole, single-throw, with enclosed fuses. Ampere-hour meter connected in battery discharge and charge circuit. Ammeter connected in battery discharge and charge circuit. Pilot-light socket connected across generator leads. Automatic switch connected in generator circuit, automatically closes generator circuit to line when generator voltage rises to proper value, and opens on small reverse current. Generator field rheostat furnished only as extra and to purchaser's specifications. Standard board includes provision for mounting front of board type of rheostat. Ground-detector lamp sockets for testing for grounds.

Utility of Ampere-Hour Meter.—The ampere-hour meter, which is a feature of these switchboards, is of great practical value. It is to the storage battery what a tank gauge is to a water tank. It indicates at all times the current in ampere-hours taken out of the battery. The hand moves from "Full" toward "Empty" when the battery is discharging, and from "Empty" toward "Full" when

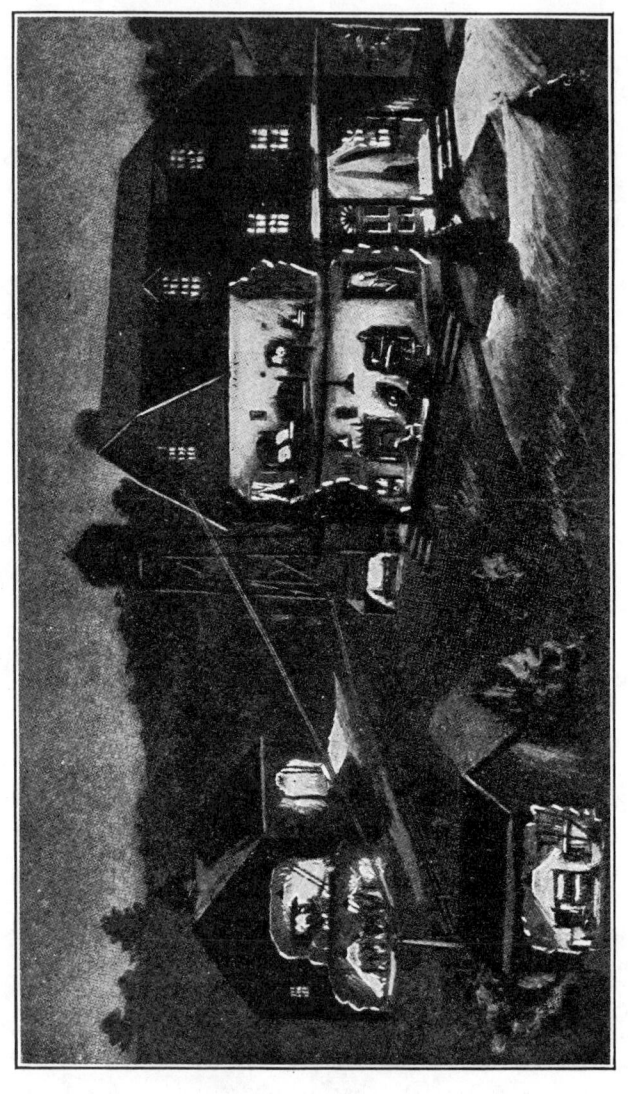

Fig. 70.—Showing Adaptability of Electric Lighting From an Isolated Plant for Farm Use.

UTILITY OF AMPERE-HOUR METER 153

the battery is charging. It is arranged to run somewhat slow on charge, and thus automatically provides for the necessary excess of ampere-hours of charge over ampere-hours of discharge. It is provided with a contact at the "Full" point, of which use may be made to light a lamp or ring a bell when the meter hand reaches that point, or to actuate a circuit-opening device in the

Fig. 71.—Isolated Lighting Plant Installed in House Basement Using Edison Battery. Engine Power is Used to Drive Water Pump, Washing Machine and Other Domestic Machinery.

engine ignition circuit, shutting the engine down when the hand reaches the "Full" point. The ampere-hour meter is used:

1. To show when the battery needs to be charged.
2. While charging, to show at what rate the charging may be done, permitting the use of relatively high charging rates and thus shortening very materially the hours of engine operation for battery-charging purposes.

154   STORAGE BATTERIES SIMPLIFIED

3. To show when the battery has been fully charged and charging may be discontinued.

4. As a positive safeguard (if the simple instructions are followed) against useless, wasteful and injurious overcharging or overdischarging of the battery.

5. As a continual and valuable check on the battery performance, and therefore on battery conditions.

6. To measure the daily or weekly current consumption, or the current used by flatirons, motors or other devices using current intermittently in the performance of some specific task, and to

Fig. 72.—Sectional View of Delco-Lite Gasoline Motor and Dynamo, Showing All Parts of Power-Producing and Current-Generating Units.

UTILITY OF AMPERE-HOUR METER

obtain other data of like character, which are helpful in operating the plant with economy, in preventing the waste of current, etc.

7. To give, when such is desired, a signal by light or bell, that the battery charge has been completed and should be discontinued, or to actuate a stop-charge device.

Delco-Lite Outfit.—The system shown at Figs. 72 and 73 was developed by one of the largest producers of lighting systems for automobiles, and incorporates many of the features of simplicity and ease of operation that have been essential in automobile work. This combined, compact unit consists of a gasoline engine, a dynamo and a switchboard. It weighs about 325 pounds, exclusive of the storage batteries, which are furnished as a part of the outfit. It is a compact plant that will deliver 750 watts. It is a low-voltage system—32 volts—as this saves battery expense; at the same time it is of sufficient voltage to operate light machinery efficiently. It is now possible to purchase almost anywhere standard motors and lighting fixtures for this voltage. There is not the slightest danger in handling this low voltage.

The gasoline engine is of the air-cooled type, so there is no danger of freezing, no matter where the outfit may be located. It is self-starting. All that is required is to close the switch, which starts the engine. It automatically cuts off when the batteries are charged.

The batteries are of the sealed glass-jar type, especially built for use with this outfit, and come fully charged. It is said that they will not freeze at 20 degrees below zero, even when completely discharged. Extra large space is provided for electrolyte or liquid, and they are long-lived.

Any number of lights may be installed, up to 50 or 60. The average place, however, burns only a few of these at any one time. When the engine is running it will carry thirty-two 20-watt lights continuously. The storage battery alone will carry fifteen 20-watt lights for eight hours. Of course, increased storage capacity can be provided. The batteries constitute a reserve source of energy, providing current that may be used when a light is turned on or if some light machinery be operated, such as the churn, washing-machine, cream separator or vacuum cleaner. The engine need

156   STORAGE BATTERIES SIMPLIFIED

only be run at intervals of perhaps once or twice a week, depending upon the amount of current used. The claim is made that the average farm can be lighted for less than five cents per day. This is less than half the rate in most cities.

Fig. 73.—How the Delco-Lite Generating Outfit is Coupled to the Storage Battery.

Storage Batteries in Electric Train Lighting.—The development of practical train lighting by electricity was a great step forward, and its advantages were thoroughly appreciated by the public. Electric light contributes to the safety of the traveling public, as it lowers the fire risk present with either oil lamps or gas flames. Any practical storage cell may be used for train-lighting service, as satisfactory installations have been made with either the lead-plate or Edison batteries. This service is not exceptionally severe, as an examination of some lead-plate batteries with a record of 500,000 car miles during a period of three years showed that the jars contained but one inch of sediment, and plates were in excellent condition. The results point to a normal life of

Fig. 74.—Edison Storage Battery for Train Lighting, Showing Arrangement of Cells in Trays of Three, to Facilitate Handling.

ten years with but one intermediate cleaning. The batteries used are very similar in general construction to those intended for use in electric vehicles, and are installed in trays for ready handling. The illustrations at Fig. 74 show the application of the Edison cell to this service and the method of grouping the cells to form a battery. The location of the battery compartment on the car and accessibility of the battery trays are clearly depicted.

158 STORAGE BATTERIES SIMPLIFIED

While the battery is an important part of a train-lighting system, the dynamo and method of regulation are also of interest. The E. S. B. system is shown in diagram form at Fig. 75, and the distinctive features making for automatic operation may be understood by a careful study of the diagram. The dynamo has a bipolar armature rotating between heavy pole shoes, each pole-piece being securely attached to the core pieces. Two pairs of brushes are used, these being spaced on quarters of the commutator, or 90 degrees apart. One pair is short-circuited, while the other is coupled through the series winding to the outside circuit. These brushes are known as the "short-circuit brushes" and the "load brushes," respectively. The pole pieces are provided with the usual field coils, F^1 and F^2. The series winding F^2 is connected between the generator terminal and the top load brush. The control field winding, F^1, is connected across a Wheatstone bridge W and provides the primary field excitation. The magnetic field produced by this primary excitation passes across the armature in the direction of the arrow P, and then through pole shoes, pole necks and frame of the machine. This magnetic field is, under normal operating conditions, of very small strength, producing a low voltage between the short-circuit brushes B^1. This low voltage, however, produces a sufficient flow of current through the short circuit between these brushes and through the armature winding to develop a considerable magneto-motive force. This latter magneto-motive force produces a magnetic field at right angles to the primary field, which passes through the armature and pole shoes, as shown by arrows K, but does not pass through the pole cores or frame of the machine. This secondary field flux produces the desired voltage across the load brushes B^2.

One of the principal results obtained with the Rosenberg type of machine is the development of the same polarity for either direction of rotation without employing any pole changer, or any alterations whatever in the circuit connections. This is due to the fact that when the direction of rotation changes the voltage generated across the short-circuit brushes by the primary field is reversed, and the current flowing between these brushes is, therefore, reversed in direction. This reverses the direction of the secondary field K, which produces the voltage across the load brushes B^2. It will be seen, therefore, that when the direction of armature rotation is reversed the direction of the main or secondary field excitation is also reversed, producing no change in polarity. The importance of this can be realized when one takes the method of driving the dynamo armature into consideration. It is belt-connected to one of the car axles and is just as apt to be driven in one direction as the other.

The E. S. B. dynamo is controlled for constant voltage rather

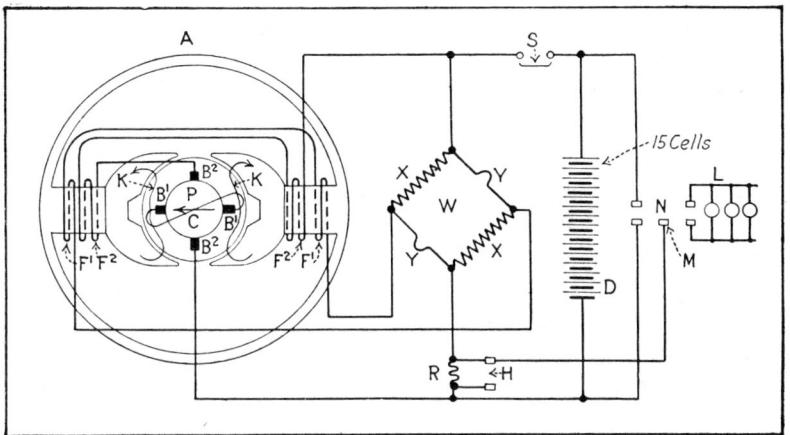

Fig. 75.—Diagram of Dynamo Used with E. S. B. Car-Lighting System.

than constant current, the voltage being held at approximately 33½ volts when used with 15-cell equipments. This control is effected by means of a Wheatstone bridge W connected across the machine terminals at two opposite junction points of the bridge, while the other two opposite points are connected to the field terminals. The Wheatstone bridge consists of two fixed resistances, XX, and two iron wire "ballast" resistances, YY. The iron wire "ballast" resistances have the characteristic of increasing their resistance rapidly with very small increments of current when operating at a dull-red heat. The design of this bridge (for 15-cell equipments) is such that when the machine voltage is $33\frac{1}{2}$ volts the resistances of X and Y are practically equal. Under these conditions no current will flow through the field winding F^1. When the voltage is lower than $33\frac{1}{2}$, the resistance of Y is less than the resistance of X, and Y will, therefore, carry more current than X, the excess current flowing through the field winding F^1. As the voltage of the machine increases, therefore, the field excitation becomes smaller and smaller, approaching zero as the voltage approaches $33\frac{1}{2}$ volts. At high speed the current in the field is actually reversed in order to partly counteract the residual magnetism in the pole necks and the frame of the machine. In series with the Wheatstone bridge is a fixed resistance R, which is normally short-circuited by a switch H, and is also short-circuited by a contact clip or by an extra blade on the lamp switch. When the switch H and the lamp switch are both open, the resistance H is no longer short-circuited. The drop in this resistance then lowers the voltage applied across the Wheatstone bridge, and in order to restore the latter to its normal balanced value, the voltage of the machine will be increased by an amount equal to the drop in the resistance R. This arrangement permits the voltage of the machine to be increased during a daylight run, when the lamps are not in use, in order to give the battery a high-voltage charge if this should ever be deemed necessary or advisable. When the lamp switch is closed the resistance R is immediately short-circuited and the voltage of the machine is restored to normal, so that the lamps are never subjected to this high voltage. On the actual switchboard, the switch H consists of two terminals connected by a brass rod, which, by sliding lengthwise, may be disconnected from one of them. While this means for increasing the voltage has been included to provide for possible contingency, it has not been found necessary to use it on any of some seventy equipments in service from two to three years.

Automatic means are provided to prevent a battery overcharge, and a "cutout" arrangement is included to prevent the battery discharging through the generator when its voltage is greater than the electromotive force of the dynamo. Means are provided to supply generator current to the battery as it is used, so the cells are always maintained in a charged condition. The features of

STORAGE BATTERIES IN TRAIN LIGHTING 161

current control developed in car-lighting service have been adopted by automobile makers in the modern starting and lighting systems which are so popular and now considered indispensable. The experience of the various storage-battery makers in producing satisfactory batteries for this service was turned to good

Fig. 76.—Types of Storage-Battery-Propelled Locomotives for Industrial and Mine Haulage.

advantage in building the battery types intended for automobile engine-starting work.

Storage-Battery Locomotives.—The conditions under which storage batteries used in mine locomotives or those intended for industrial work operate call for very careful selection and installation. The battery should always be mounted in the locomotive so the flat of the plate comes against the direction of motion. When the plates are thus assembled in the jars there is no opportunity for displacement, as would be the case with the plates arranged so the edges were facing the direction of movement. A cushioning effect is obtained by the electrolyte against the plate, which is very valuable, as the many shocks incidental to the none too gentle coupling and uncoupling of the locomotive to the loaded cars and shocks due to sudden starts are thus minimized. Rubber cell jars should be considerably heavier for this service than are ordinarily used in automobile work. The trays in which the cells are carried should be provided with substantial partitions so the jars at the ends of the trays will not be subjected to the inertia of the remaining cells. Inasmuch as metal enters so largely into the construction of locomotives, the battery trays should be thoroughly insulated from the frame and should be securely blocked in place so there will be no possibility of shifting. The battery should be mounted apart from the propelling machinery or motor, and so mounted on the frame to permit drainage of the battery compartment. The illustrations at Fig. 76 show typical locomotives for mine and industrial use, and in one the sides of the battery compartment are dropped to show how the battery is mounted and how accessible the trays are.

In case it is desired to make a rough approximation of the size of battery needed, the Electric Storage Battery Company gives a method that is very simple, and while the figures are approximate, they enable the engineer to determine the type and size of battery best adapted for the individual requirements. If boosting charges can be given then batteries of lesser capacity can be used than if the machine must operate for more extended periods on one charge. The number of cells used will, of course, vary with the type of locomotive and its weight. A three-ton locomotive can be operated on 48 MVII-type cells, and by increasing the capacity of the cells by making them larger the same number can be used for a twenty-five-ton locomotive for freight-yard use and shifting loaded freight cars over tracks in city streets during the night hours when the traffic is light. Of course, the charging current available must be taken into consideration. A 48-cell battery can be charged directly from a 110-volt direct current. In some cases, where a power circuit of 220 volts direct current is used, more cells may be provided and charged directly from the power line. An example of this kind is an eight-ton mine locomotive equipped with 100 MV19-type "Iron Clad" Exide cells.

How to Figure Mine Locomotive Battery Capacity.—The figure of 30 pounds per ton, which is taken as the tractive effort required to overcome friction, is considered sufficient to include an allowance for curves which may be encountered as well as some allowance for tracks that are not up to good railroad standard. The efficiency of $66\frac{2}{3}$ per cent. assumed between the battery and locomotive wheels includes an allowance for rheostatic and braking losses as well as for motor and gear losses. Different values may be assigned these constants, as the judgment of the engineer making the approximate calculations may dictate, without changing the method of figuring. The typical round trip should be divided into as many parts as have different characteristics, such as different weight of train and different grades to be encountered. The requirements of each of these sections in each direction of haul should be treated separately in the following manner:

(a) Determine the tractive effort required for level running by multiplying the weight of the train in tons, including locomotive by 30 pounds per ton to get the tractive effort to overcome friction on the level.

(b) Find the tractive effort for grade, if any is encountered, by multiplying the grade expressed in per cent. by 20 and multiplying this by the weight of the train in tons, taking this figure as positive for up grade and negative for down grade.

(c) Add the tractive effort level running and the tractive effort for grade to get the total tractive effort. If the sum is negative, it means the train is coasting and no power is required.

(d) Multiply the total tractive effort by 3 to get watt-hours battery per train mile. (1 watt-hour equals 2.655 foot-pounds, which is very nearly ½ a mile per pound; tractive effort $\times$ 2 = watt-hours per train mile at the locomotive wheels. Assuming $66\frac{2}{3}$ per cent. efficiency between the battery and the wheels, we have

$$\frac{\text{tractive effort} \times 2}{.66\frac{2}{3}}$$

= tractive effort $\times$ 3 = watt-hours per train mile.)

(e) Multiply the watt-hours per train mile by the number of miles of total operation on this section of the track in this direction to get the watt-hours required for this portion of the operation. Repeat the above for all sections of the track and in both directions.

(f) Add the watt-hours required for all portions of the operation, and the sum of these gives the total watt-hours of battery capacity required.

(g) Divide the total watt-hours by the voltage of the battery (the number of cells multiplied by 2) to get the ampere-hours of battery capacity.

(h) Divide the ampere-hours of battery capacity by 31.5 to get the number of positive plates per cell. The figure 31.5, which is the 4½-hour capacity of a positive plate in ampere-hours, can only be used when there are at least 5 cycles of operation approximately evenly distributed over 4½ hours or longer.

(i) Multiply the number of positive plates by 2 and add 1 to get the number of plates per cell.

The example which follows considers only one case, and, of course, applies only to the conditions stated. Each individual application must be considered with full realization of the conditions obtaining, but the procedure to be followed is substantially the same as outlined in any case. In any event, it is always well to consult the engineering department of the battery maker before using batteries for any purpose, as this results in securing advice that will assure a successful installation.

Example.—Assume 600 feet level track and 800 feet of 1.2 per cent. grade, a 3-ton locomotive, a 15-ton trailing load in the

EXAMPLE

direction against the grade and 6 tons trailing load in the direction with the grade, 20 round trips per 8-hour day, 48-cell battery.

This round trip can be considered in three parts:

First, 800 feet up 1.2 per cent. grade, 18-ton train.

Second, 800 feet down 1.2 per cent. grade 9-ton train.

Third, all level running can be grouped, using the full distance and the average load 1,200 feet level, 13.5-ton train.

FIRST PART

(a)
```
    18   tons
    30   pounds per ton friction
   ───
   540   pounds TE for friction
```

(b)
```
   1.2   per cent. grade
    20   pounds per cent. per ton
   ───
    24   pounds per ton grade
    18   tons
   ───
   432   pounds TE for grade
```

(c)
```
   540   pounds TE for friction (a)
   ───
   972   pounds TE total
```

(d)
```
     3
   ───
  2916   watt-hours per mile
```

(e)
```
           800 feet per trip
            20 trips
           ───
    5280 | 16000 feet
           3.03 miles

    2916 watt-hours per mile (d)
    3.03 miles
    ───
    8830 watt-hours for first part
```

Second Part

(a) $\begin{cases} 9 \text{ tons} \\ 30 \text{ pounds per ton friction} \\ \hline 270 \text{ pounds TE for friction} \end{cases}$

(b) $\begin{cases} -24 \text{ tons per ton for grade } (b, \text{ first part}) \\ 9 \text{ tons} \\ \hline -216 \text{ pounds TE for grade} \end{cases}$

(c) $\begin{cases} 0 \text{ pounds TE for friction} \\ -216 \text{ pounds TE for grade} \\ \hline 54 \text{ pounds TE total} \end{cases}$

(d) $\begin{cases} 3 \\ \hline 162 \text{ watt-hours per mile} \end{cases}$

(e) $\begin{cases} 3.03 \text{ miles } (e, \text{ first part}) \\ \hline 491 \text{ watt-hours for second part} \end{cases}$

Third Part

(a) $\begin{cases} 13.5 \text{ tons} \\ 30 \text{ pounds per ton friction} \\ \hline 405 \text{ pounds TE for friction} \end{cases}$

(b)   No grade

(c)   405 pounds per ton total

(d) $\begin{cases} 3 \\ \hline 1215 \text{ watt-hours per mile} \end{cases}$

(e)
$$\begin{array}{l} 1200 \text{ feet per trip} \\ 20 \text{ trips} \\ \hline 5280 | 24000 \text{ feet} \\ 4.55 \text{ miles} \\ 1215 \text{ watt-hours per mile} \\ \hline 5530 \text{ watt-hours, third part} \end{array}$$

(f)
$$\begin{array}{l} 491 \text{ watt-hours, second part} \\ 8830 \text{ watt-hours, first part} \\ \hline 14851 \text{ watt-hours, total} \end{array}$$

(g) 48 cells at 2 volts per cell gives 96 volts
96 | 14851 watt-hours

(h) 31.5 | 155 ampere-hours
4.7 positive plates, 5 must be used

(i) $5 \times 2 = 10 \quad 10 + 1 = 11$
Battery will be 48 cells, MV-11.

Storage-Battery Street-Railway Car.—The first practical power-propelled street-cars were converted horse-drawn types electrified by the use of storage batteries. The defects of the early storage-battery designs and the development of the present overhead feed-wire and trolley, as well as the utilization of high-voltage motors for this purpose, halted development of battery cars for a period. The improvements that have been made in modern storage batteries show that it can now occupy a position of importance in the transportation field. A variety of systems of street-car propulsion are available, each having a distinct field of usefulness. They may be divided into broad classes, one whose units are supplied with power from a central station, with which they must always be in contact, and those employing self-contained units. In the first class, we have third-rail, cable-conduit and trolley systems; in the second are grouped storage-battery and gasoline-electric types. The Electric Storage Battery Company have made a study of the battery car problem, and with the co-operation of the car builder and electrical apparatus manufacturer, types have been designed that are especially adapted for street-car work. Cars of very high speed and power have been successfully propelled by storage-battery current, as well as the single-truck cars used in belt-line service.

Among the situations which may present conditions favorable to the use of storage-battery cars are the following:

1. Short rural lines not connected with other systems, yet meeting a distinct transportation need.

2. Extensions or spurs to existing electric systems, serving districts which it is desirable to develop, but where the immediate traffic will not justify overhead construction investment.

3. Lines operated to meet some local and special transportation demand other than general public service, as, for example, between mills or factories and a main railroad depot or residential center, plantation railroads, belt-line service in large manufacturing plants, and the like.

4. For the operation of branch lines or extensions used only during certain seasons of the year, as in connection with seaside and summer resorts, amusement parks and the like.

5. As an adjunct to steam lines to supplement the regular steam-train service, furnish local service at short intervals and to branch lines; particularly on roads devoted largely to freight traffic, where there is need of a passenger service, but insufficient traffic to justify regular steam-train operation.

6. Where local ordinances or other reasons do not permit the use of overhead trolley construction.

7. For providing infrequent night or "owl" service.

The storage battery for a single-truck car is located beneath longitudinal car seats in ventilated compartments, and can be easily reached for inspection by removing the seats and thus exposing the battery cell tops. It is composed of 58 cells, of type MV29 Exide, having a capacity of 67 amperes for six hours' continuous discharge at an average of 114 volts. The cells are contained in substantial wooden trays, each containing four cells. Fourteen such trays are provided, the total number of cells

STORAGE-BATTERY STREET-RAILWAY CAR 169

mounted in this manner being fifty-six. Two extra cells are needed to bring the battery up to the desired capacity, these being carried in smaller individual trays. The trays are convenient units for easily removing the battery from the car if necessary, but as it may be charged in place and electrolyte readings and evaporation loss compensated for by removing the seats, in ordi-

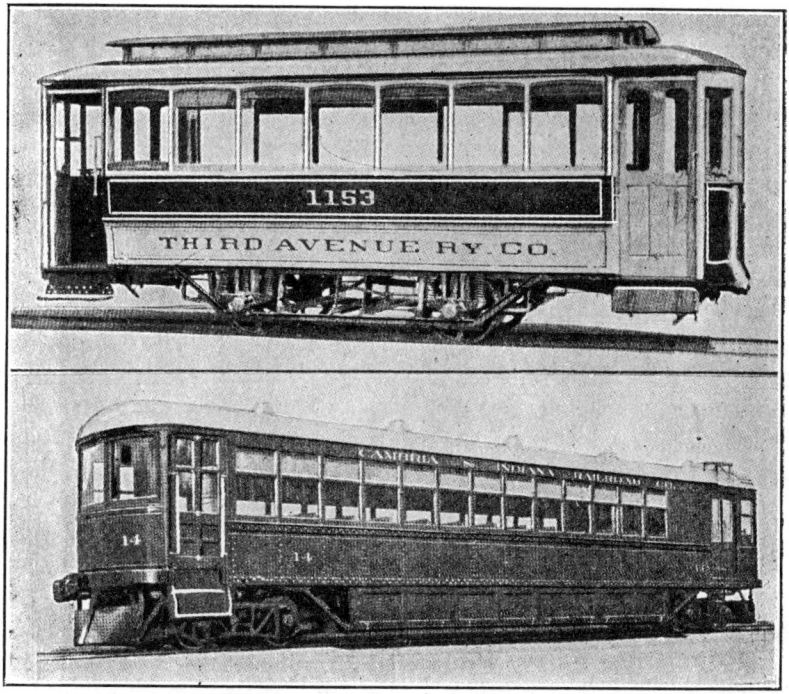

Fig. 77.—Storage-Battery-Propelled Street-Railway Cars. At Top—Single-Truck Car for Belt-Line Work in Cities. Below it, Fast Double-Truck Interurban Car.

nary operation the battery need not be removed from the car. The total weight of the battery is 4,876 pounds. The cells are practically of the same general construction as heavy motor-truck batteries, being of the sealed type. Cars equipped with a battery of this capacity will take a 7 per cent. grade with full passenger load and attain speeds of 15 miles per hour on level track. Under average operating conditions, the input to the car motors, which are of the automobile type, will average 450 watt-hours per car mile. At this rate a car will make 100 miles on a single battery charge on the basis of the six-hour battery capacity specified. In actual service the cars have been run 120 miles per charge. The total weight of such a car is 14,000 pounds, including battery. The seating capacity is sufficient for 26 persons, and the car is 18 feet long.

Larger cars, 28 feet long, of the two-truck type, have a seating capacity for 36 persons. The battery for such cars is 88 cells, of type MV29 Exide battery having a capacity of 67 amperes for six hours' continuous discharge at an average of 173 volts. The battery is contained in 44 trays of 2 cells each. The battery weight is 7,392 pounds, and total weight of car, including battery, is 13 tons. The maximum speed possible is 25 miles per hour, and the current consumption is about 700 watt-hours per car mile. The range of action is 80 miles on one charge. As is true of the smaller car, the battery is carried under the longitudinal seats. In some larger cars, such as shown at the bottom of Fig. 77, the batteries are mounted under the car floor similar to a train-lighting battery, and occupy all the available space between the trucks. When this method of installation is followed the cells are mounted in easily handled trays, as in lighting service.

Submarine Boat Batteries.—One of the most spectacular applications of the storage battery is to submarine boat propulsion when these are under water. On the surface, the craft is propelled by Diesel type internal-combustion engines, which deliver part of their power to generators, which keep the batteries charged. Of course, when submerged, the internal-combustion engines must be shut off and the stored energy of the battery drawn on to drive the boat; the generator becomes a motor for ship propulsion and derives power from the battery it had previously charged. In addition to the main driving motors, there are several others for pump actuation, steering, etc., all of which use battery current. A sectional view of a typical submarine using the Edison alkaline

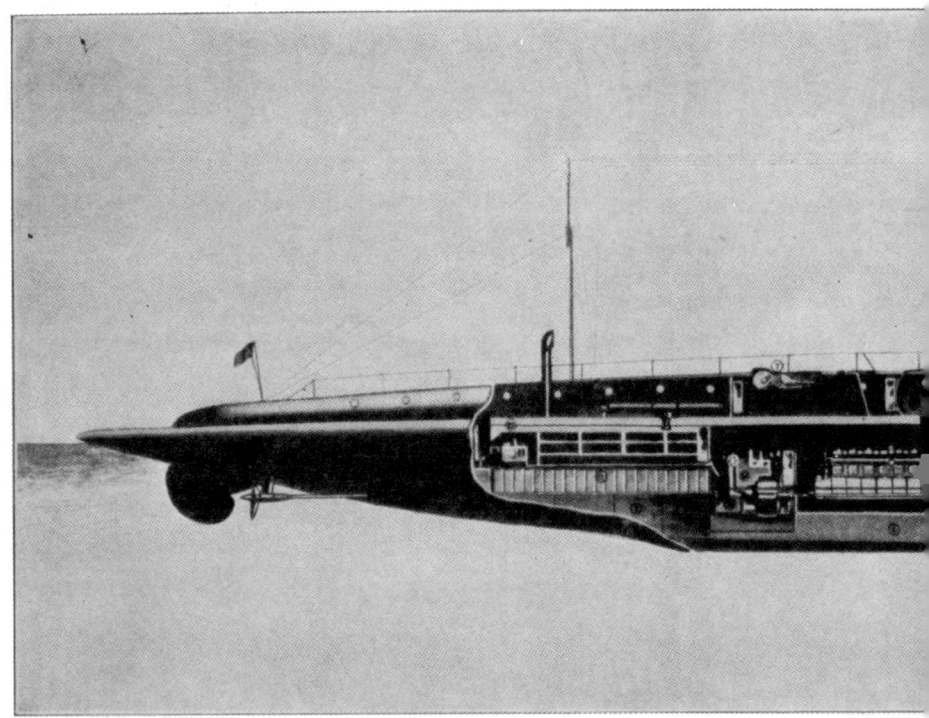

Fig. 78.—Sectional View of Lake High-Speed, Sea-Keeping Fleet Submarine. 1—Main Ballast Tanks. 2—Fuel Tanks. 3—Keel. 4—Safety Drop Keel. 5—Habitable Superstructure. 6—Escape and Safety Chambers. 7—Disappearing Anti-Aircraft Gun. 8—Rapid-Fire Gun. 9—Torpedo Tubes. 10—Torpedoes. 11—Twin Deck Torpedo Tubes. 12—Torpedo Firing Tank. 13—Anchor. 14—Periscope. 15—Wireless. 16—Crew's Quarters. 17—Officers' Quarters. 18—Warhead Stowage. 19—Torpedo Hatch. 20—Diving Chamber. 21—Storage Battery. 22—Galley. 23—Steering Gear. 24—Binnacle. 25—Searchlight. 26—Conning Tower. 27—Diving Station. 28—Control Tank. 29—Compressed-Air Flasks. 30—Forward Engine-Room and Engines. 31—After Engine-Room and Engines. 32—Central Control Compartment. 33—Torpedo Room. 34—Electric Motor Room. 35—Switchboard. 36—Ballast Pump. 37—Auxiliary Machinery Room. 38—Hydroplane. 39—Vertical Rudders. 40—Signal Mast.

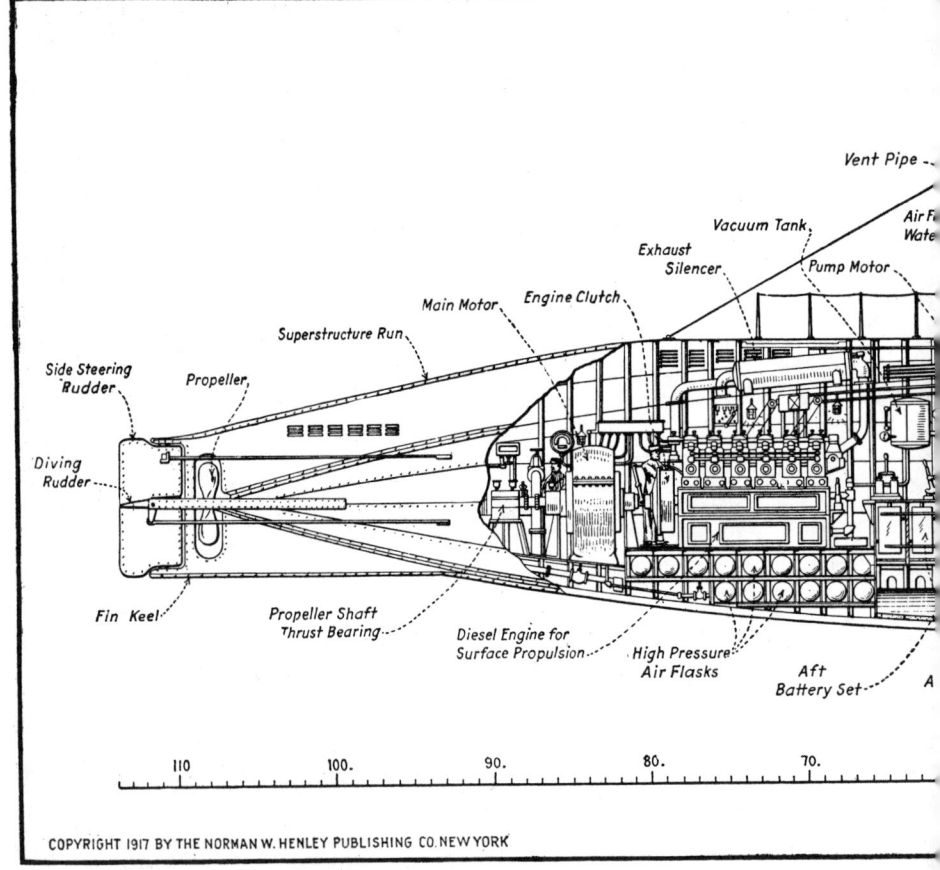

Fig. 79.—Sectional View of Typical Modern Submarine Boat, Showing Disposition of Propulsive Machinery for Driving Boat on the Surface and Arrangement of Batteries for Underwater Power.

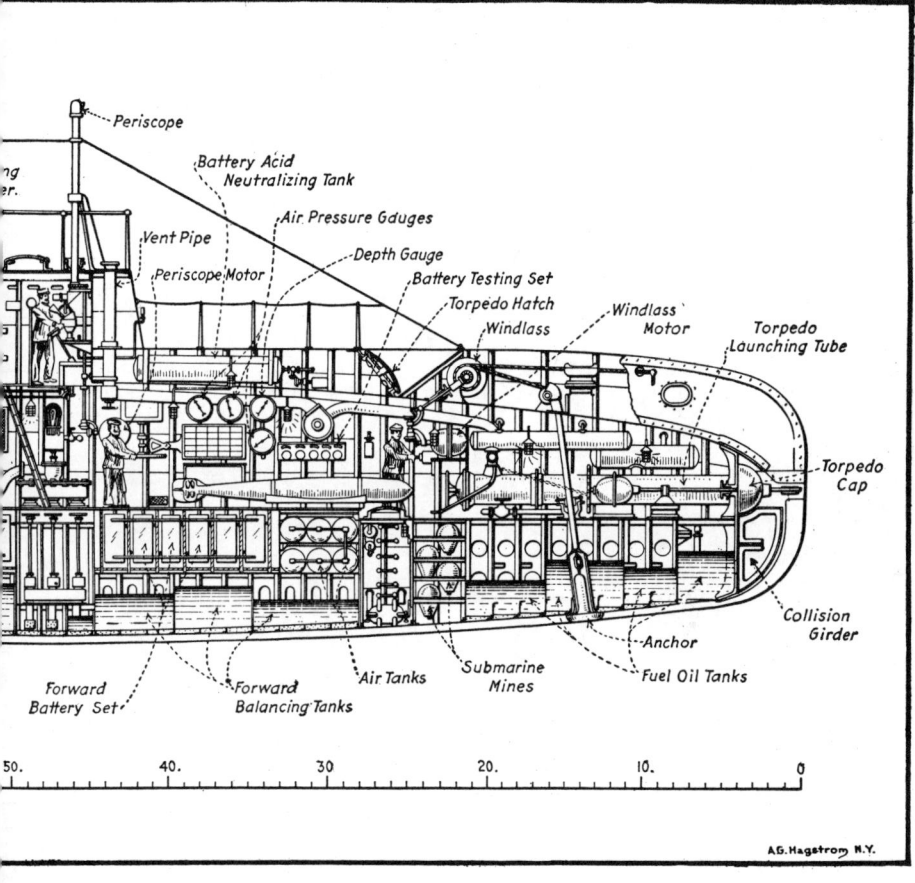

SUBMARINE BOAT BATTERIES

battery is shown at Fig. 78 and one using batteries of the lead-plate type is depicted at Fig. 79. The details of the battery placing and location of propelling machinery are sufficiently clear so the general details may be easily understood.

There seems to be considerable difference of opinion relative to the best type of storage battery for submarine work, the advocates of the alkaline battery offering a number of arguments that merit consideration. A typical submarine boat battery cell of the alkaline type is shown at Fig. 80 A, the negative plate at B and the positive plate at C. The general structure is the same as the smaller Edison cells, except that the plates are built up of a number of sub-grids that correspond to the elements of the smaller batteries, riveted to a main frame of sheet steel to form plates of the required capacity. The cells are installed so that they will remain securely in place regardless of the shaking up they receive when the boat is in rough water. The cells are constructed so no electrolyte can escape or water enter. The Edison battery is also built strong enough so gas explosions can take place in its interior without damaging the parts.

The illustration at Fig. 78 was furnished by the Edison Storage Battery Company, and shows a proposed design of Lake Torpedo Boat fleet submarine, in which one can see the location of the battery, indicated by "21." It would necessarily be quite a large battery, having a capacity of about 10,000 ampere-hours, at 220 volts.

The sectional view at Fig. 81 is a typical coast-defense type of submarine. As described by M. R. Hutchison, E.E., engineering advisor to Thomas A. Edison, the details are: "This is a smaller boat, and is 135 feet long, 14 feet beam. The displacement submerged, 337 tons. Engines, two in number, are about 300 horse-power each. The two propelling motors are about 115 horse-power each, 220 volts. The surface speed is about thirteen knots, submerged speed about nine and one-half knots. The battery would consist of two hundred Edison cells, the total number in each of the tanks being one hundred. The drawing will show the location of the battery tank. The capacity of such a battery would be about 900 kilowatt-hours at the three-hour rate of discharge, and about 540 kilowatt-hours at the one-hour rate of discharge. This capacity would be in excess of the capacity of any lead storage battery installed in the same tank, and would be five or six tons lighter than a lead-sulphuric-acid battery installed in the same tank. This will naturally give the vessel an increased cruising area under water and provide excess capacity, because of the reduction in weight, for fuel oil."

The advantages of the Edison battery are given by the same authority as follows:

"First, it is the lightest battery, and occupies the least space for a given output of any now obtainable.

"Second, its electrolyte is non-corrosive, is not injurious to the steel work of the hold or fittings of the boat.

"Third, it cannot cause leaks by corrosion of rivets or joints, no matter how much electrolyte might escape from the cells.

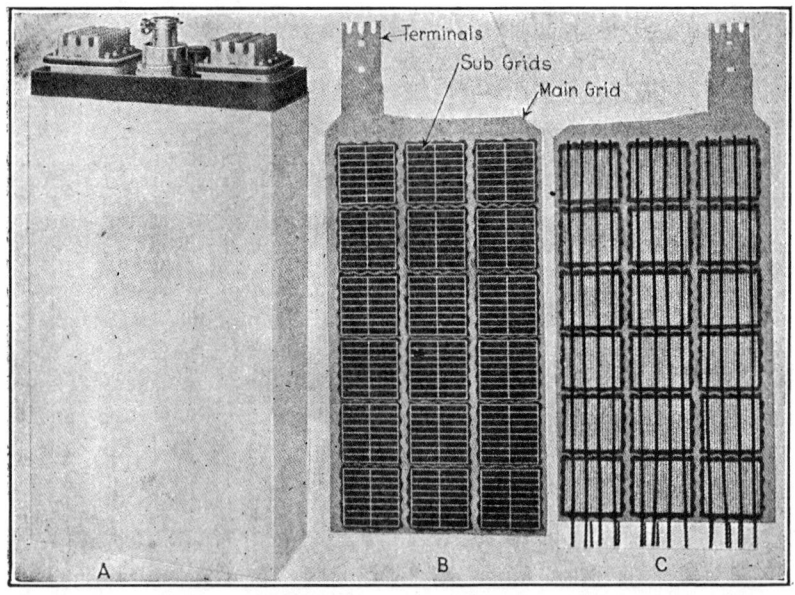

Fig. 80.—The Edison Submarine Boat Battery and Two of the Plates Showing Use of Sub-Grids.

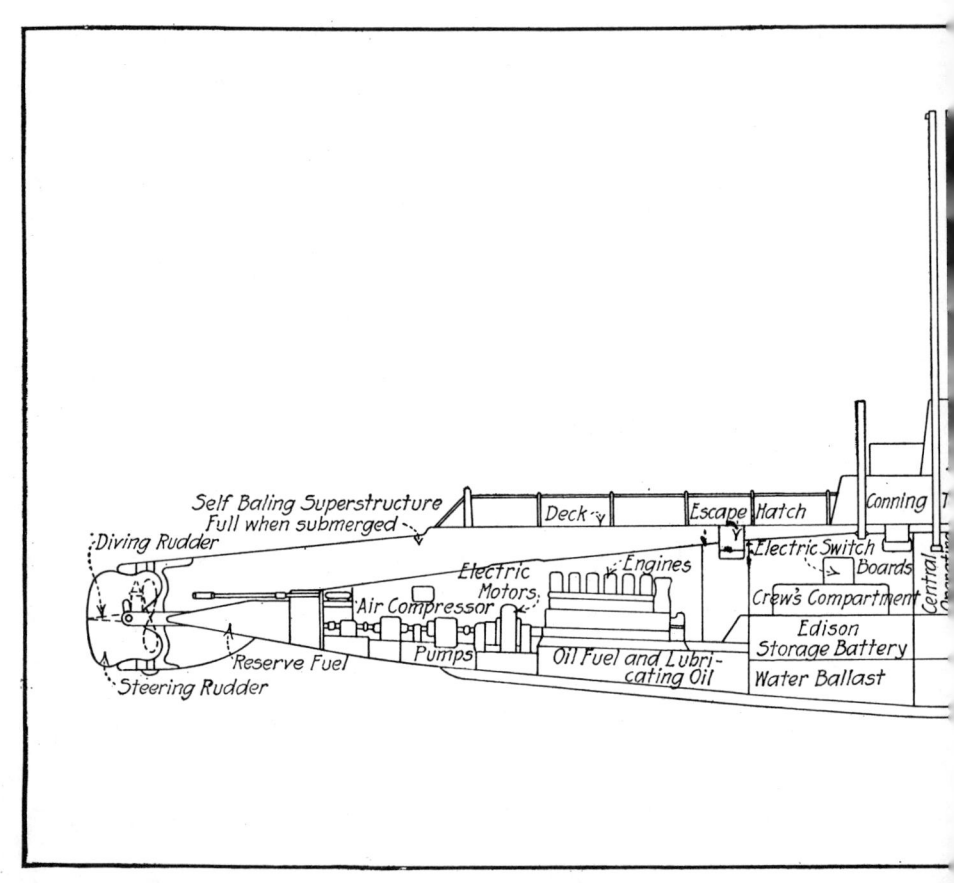

Fig. 81.—Typical Coast Defense Type Submarine Using Edison Storage Battery

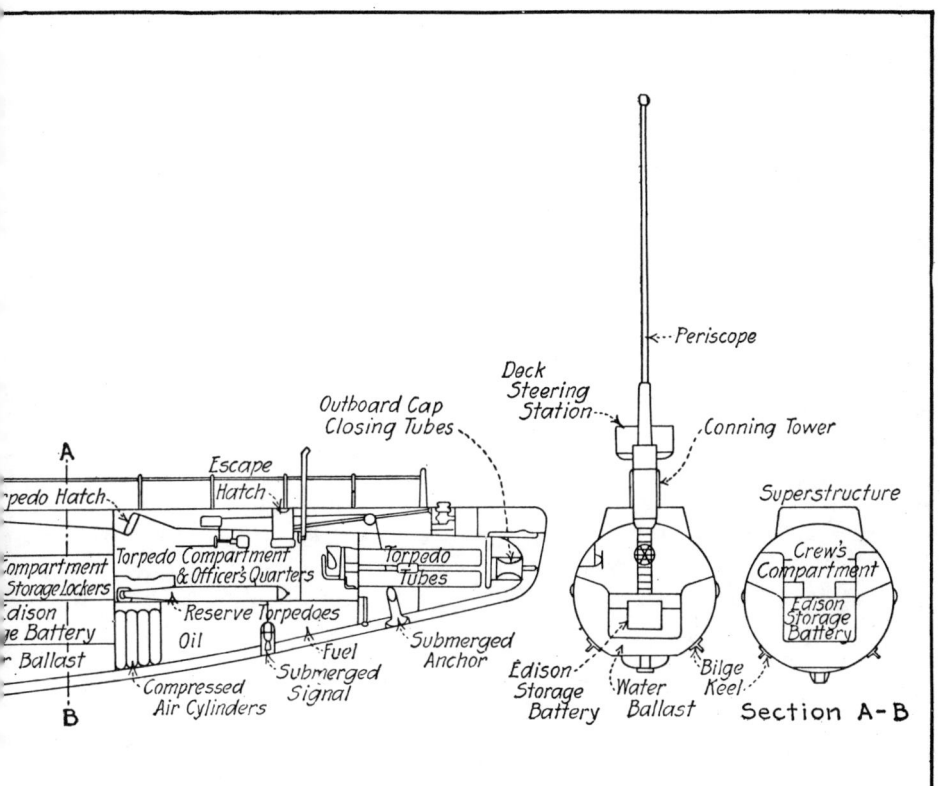

"Fourth, in the event of sea-water entering the boat, through leaks caused by external explosion of submarine mines, etc., the personnel cannot be asphyxiated by the formation of chlorine gas.

"Fifth, its mechanical structure makes it capable of resisting shocks, rough treatment and concussion from nearby external explosion of submarine mines and bombs.

"Sixth, it is impossible to buckle or distort the plates by rapid charge or discharge.

"Seventh, the active materials do not shed off with use.

"Eighth, the capacity does not diminish with use until the life of the battery is nearly exhausted.

"Ninth, it does not require dismantling to remove sediment.

"Tenth, it has a longer life under any given conditions than any other available battery.

"Eleventh, the individual cells are comparatively light, and may be easily handled if necessary.

"Twelfth, an excess emergency capacity may be stored by prolonged overcharge.

"Thirteenth, the plates are not injured by standing discharged in the electrolyte for any length of time.

"Fourteenth, it is unnecessary to remove plates or electrolyte when a boat is laid up or put in reserve.

"Fifteenth, no hydrometer readings are necessary.

"Sixteenth, it is not necessary to carry a supply of electrolyte.

"Seventeenth, it can be charged at high rate. In fact, a full charge can be put into the battery in one hour if the electrical machinery is available for doing this.

"Eighteenth, it is never necessary to remove sediment or replace separators, and also because of its long life a boat equipped with an Edison storage battery can be kept in service and does not have to be laid up on account of battery conditions when it may be required for military purposes.

"Nineteenth, in an emergency sea-water can be used for the replenishment of electrolyte. While if this is done frequently the plates will ultimately lose some capacity, no harm results from doing it a few times under stress of military necessity.

"Twentieth, oxygen is given off at a low rate of discharge.

Just enough to maintain the atmosphere of the boat in condition for the crew in any long period of submergence.

"Twenty-first, carbon dioxide is absorbed by the potash of electrolyte if the air can come in contact with the liquid. In case of protracted submergence, air can be circulated through the electrolyte by passing it through the drain tube and allowing it to emerge from the filler opening. While this is not a function of the battery, the benefit to clear the atmosphere of carbonic acid gas may prove of considerable value in case of enforced protracted submergence.

"Twenty-second, the plates cannot be injured by rubbing together when the boat is in a seaway, and plate necks cannot be broken off.

"Twenty-third, it cannot generate any explosive gases except when charging and ventilation is outbound. It cannot generate an explosive mixture of gases in the boat under practical operating submerged condition, and is, for the reason that it does not generate any gases whatever on discharge, the safest battery to be used in submarines."

The Gould Storage Battery Company, through their chief engineer, A. S. Hubbard, gives the following information relative to the Gould type batteries they have supplied to foreign governments:

"The foreign batteries consist of 120 cells, 25 plates per cell, pasted positives, pasted negatives; the plates are $15\frac{1}{4}$ inches wide, $\frac{1}{4}$ inch thick, 30 inches deep. The capacity is about 2,340 amperes for one hour per cell, giving an output for the 120 cells of about 475 kilowatts for one hour. These cells weigh about 950 pounds each, complete. The cells are individually ventilated by an exhaust blower, and the gases diluted so as to avoid the danger of explosion. The diluted gases are discharged into the atmosphere when the boat is on the surface, and the batteries are being charged and are discharged into the living compartment when the boat is submerged and the batteries discharging. It should be understood, of course, that the gases given off both on charge and discharge consist of oxygen and hydrogen, on discharge mainly of the latter, and very little of it at that, and fur-

SUBMARINE BOAT BATTERIES

ther hydrogen in small quantities has no physiological effect. The cells are sealed except at the air inlets and exits to keep dirt and salt water out. The design of the cell permits of an inclination of 40 degrees without spilling the solution."

It is only fair to the lead-plate type of battery to state that many of the troubles ascribed to their use could have been eliminated by careful attention to details of installation, and that there are numerous successful installations in United States Government craft that we are not permitted to describe or illustrate. Enough has been given, however, to show how important the

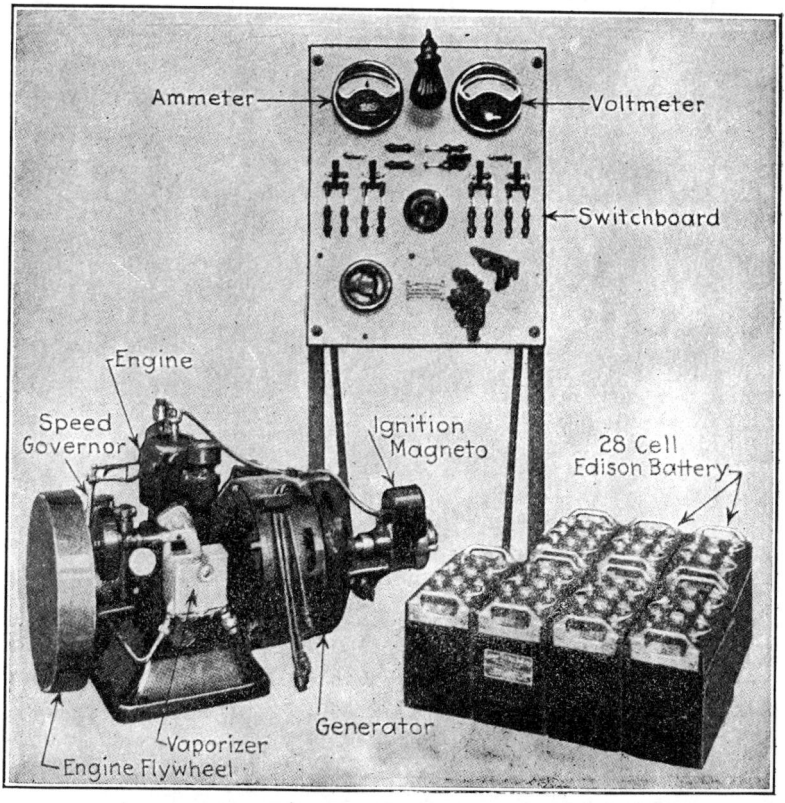

Fig. 82.—Yacht-Lighting Outfit, With Unit Power Plant, Switchboard and 28-Cell Battery.

storage battery is when the submarine is lurking in the ocean depths where no other power but electricity could be used advantageously.

Miscellaneous Marine Applications.—Storage batteries are used in many ways on shipboard, and form an essential part of the electrical equipment on a great variety of craft. Owing to the ease of installation, electric-lighting outfits are now available that will function properly in everything from the 30-foot cabin cruiser

EDISON STORAGE BATTERIES FOR YACHT-LIGHTING PLANTS
APPROXIMATE EQUIPMENT BASED ON 10-HOUR SERVICE

Approximate Length of Boat	Voltage	Number of Cells	Type of Cell	Lamps	
				Number	Candle Power
18 ft. to 30 ft............	6	5	B2 B4 B6	6 12 18	6 6 6
30 ft. to 45 ft............	12 to 20	10–20	B4 B6 A4	12 to 18 18 to 24 24 to 30	10 10 10
50 ft. to 75 ft............	30	28	B4 B6 A4	26 32 40	10 10 10
90 ft. to 300 ft...........	110	100	B2 to A12	30 to 200	16

NOTE:—Where 60-volt or 80-volt systems are required, 55 cells and 75 cells, respectively, are recommended.

to the luxurious private yacht. While power-driven outfits are available in which a dynamo is driven by auxiliary steam or gasoline engine distinct from the main power plant used in propelling the vessel, lights may be required at any time during the day or night, so there must be some constant source of current supply. For pleasure craft the continuous operation of an engine and dynamo is inconvenient and often disagreeable, and it is almost

MISCELLANEOUS MARINE APPLICATIONS 177

imperative to operate the electric lights from a storage battery charged at convenient intervals. A typical yacht-lighting outfit is shown at Fig. 82; this does not differ to any extent from the small isolated lighting plants sold for house and farm lighting. A 28-cell Edison storage battery is used in connection with the generating unit, which includes the gasoline engine and dynamo coupled together and fastened to a common base. According to the table

Fig. 83.—Electric Launch, With Storage Batteries Under the Floor.

given below, which has been furnished by the Edison Storage Battery Company, the outfit illustrated is suitable for boats varying from 50 to 75 feet in length.

Wherever charging facilities are available, the electric launch is an ideal family boat. They are noiseless and simple to control, and can be operated by anyone, even without mechanical experience. It glides gracefully and smoothly along and runs without vibration. There is absolutely no danger of fire, as is present in steam launches and to a less degree with gasoline engines. The electric motor is the simplest and most dependable source of power known, as it has but one moving part, and that is a rotating one. It is started reversed, stopped, and its speed varied by a simple wheel or lever. The battery may be easily stowed away under the floor boards, and if an Edison alkaline battery is employed, as in the launch shown at Fig. 83, the boat may be used in salt water just as well as fresh.

The alkaline battery gives absolute reliability to the electric launch by eliminating battery troubles. It is practically as rugged as the motor itself, and never disappoints the owner by going dead at critical periods. As there are no obnoxious, irritating or corrosive fumes given off by this type during charge, the charging may be done directly in the boathouse, without discomfort or the discoloration of metal trimmings on the launch or other vessels near it. During the period of idleness, which practically all pleasure craft of this kind experience, the Edison battery does not deteriorate. If charged before being laid up there will be ample power to run the launch many miles as soon as it is put into commission again. Those who have experienced the difficulties of "taking down," packing and reassembling other types of batteries will appreciate this advantage.

Reliability is of paramount importance in every detail of wireless telegraph apparatus. No doubtful or suspiciously weak device can be tolerated in a system upon which the lives of passengers and crew are likely to depend at any moment. And the auxiliary apparatus, the part that has to be ready at an instant's notice, but may be left to take care of itself for weeks and months in the absence of emergencies, must be always ready beyond the possibility of a doubt. This reliability for auxiliary service is found only in the storage battery. During periods of idleness there should be no internal deterioration or wasting away of active material. Left charged, the alkaline battery may be recharged at any time without injury, no matter how much or how little of the previous charge has been used. In the severest storms or in case of collision there is no danger of broken jars or spilled electrolyte because of the strength of the steel construction. The lighting battery of small boats may, of course, be used to operate the wireless equipment, and often the auxiliary wireless telegraph battery of larger vessels is used for a reserve lighting system, including "police" lights around the decks and cabins and emergency lamps in the running-lights. Lead-plate storage batteries have been used successfully for wireless work, but more care is needed as regards charging and discharging. Where the lighting battery is used for a wireless auxiliary, the use of the lights keeps the battery in proper condition, and the lead-plate type will give excellent service. Under conditions where the battery is apt to be neglected as regards regular charging and discharging, the alkaline type is the best.

Railway Switch and Signal Service.—An absolutely dependable source of power is needed for operating the block signals on our railways, and any failure of a semaphore or signal lamp to answer the control switch in the signal tower may result disastrously. The stored energy in a secondary battery is always available for this work, and either lead plate or alkaline batteries are suitable. The illustration at Fig. 84 shows the compact installation possible

Fig. 84.—Showing How Alkaline Battery May Be Installed in Signal Tower Adjacent to Control Switches, etc.

when batteries of the alkaline type are used. Lead batteries may be placed under a hood and the gases allowed to escape to the outside through a suitable vent if they are installed in the same room as apparatus that may be affected by fumes evolved during the charging process.

For railway signaling the alkaline battery has great advantages by reason of its electrolyte. There being no corrosive fumes or vapors produced, the batteries may be placed anywhere in the signal tower. There is no danger of injuring the most delicate apparatus, and many installations of such batteries are in the same room as and in close proximity to relays, generators, air-compressing machinery, etc.

The Stand-by Battery.—The storage battery is used in many central stations as an emergency source of current to cope with unusual current requirements or to supply electricity for a brief time in event of damage to the dynamos or their prime mover. Among some of the conditions that would call for discharge of a stand-by battery would be interruption of current supply, due, for instance, to low steam or belt breakage in a steam plant or collection of ice in the forebay of a hydraulic generating plant. A breakdown of the generating machinery, putting some of the dynamos out of commission, thus reducing the plant capacity, would necessitate the use of the emergency current. An accidental opening of a transmission line or unexpected increase of load, such as is caused by everyone turning on the electric lights because of a sudden darkening in the daytime prior to a rainstorm, will also draw on the reserve source.

In the early applications of storage batteries to central station service one of the main objects sought was to improve the daily load factor in a steam plant by discharging the battery during the evening peak or period of heavy demand and recharging it during the hours of lighter load. This will result in a marked improvement in plant efficiency because power costs less if produced at a uniform rate. Where the peak is of short duration, the cost of a battery will be less than the added steam equipment it displaces, and as it conduces to greater economy of current production, this is a clear gain. When a charged storage battery is

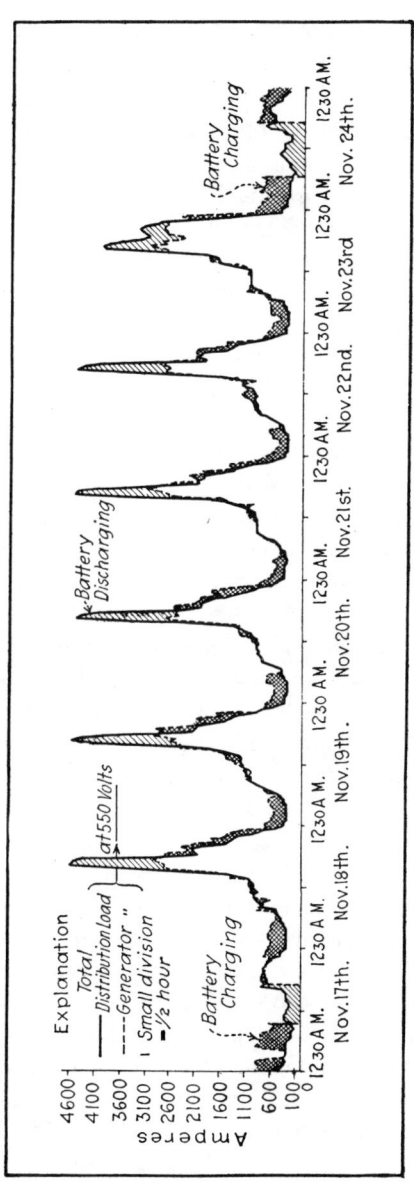

Fig. 85.—How Stand-by Battery Helps on Peak Load Work.

available as an auxiliary source of energy, it is always ready for emergency use, and this advantage is one of great value and can hardly be considered fairly on a purely pecuniary basis. The chart at Fig. 85 shows the operation of a battery in daily peak-load work. That at Fig. 86 shows how the output of a stand-by battery helped to handle an unexpected lighting load on a central station due to a particularly heavy thunderstorm. These curves are merely graphic records of the amount of current used under certain conditions and in a given time, and are easily understood.

A stand-by battery uses rugged plate construction and is always composed of very large-size cells. Type H Exide plates, which are widely used in the larger installations, are 31 inches high by 15 5/16 inches wide. The grids are castings of lead-antimony alloy and are provided with very heavy connecting lugs. When it is considered that a plate of this size may be called upon to discharge 600 amperes or more for several minutes, it will be realized that great care must be taken in proportioning the plates. The plates are hung from the cell tops by the plate lugs, which rest on vertical pieces of heavy glass arranged on either side of the tank, suitably notched to receive the plate lugs. The glass plates rest upon the reinforced lead lining at the bottom of the cell. A space of $\frac{3}{4}$ inch is left between the outside negative plate and side of the tank at one end to permit the taking of hydrometer readings. An instance of the large size of the cells is the sediment space allowed, which is 12 inches in a type H cell.

A tank suitable for an Exide element having a capacity of 3,000 amperes for one hour measures about $22\frac{5}{8}$ inches long by $21\frac{1}{2}$ inches wide. For a capacity of 6,000 amperes at the hour rate the length is increased to three feet, and for 9,000 amperes the length is nearly five feet. The height of such a cell from the floor to the busbar is about $5\frac{1}{2}$ feet, or high enough so the average man can barely look into the cell. The tanks are built of specially selected yellow pine, put together with glued, dovetailed and doweled joints. No nails or metallic fastenings of any kind are used. The lumber used is of sufficient strength to be entirely self-supporting. These tanks are treated with two coats of acid-resisting paint inside and out.

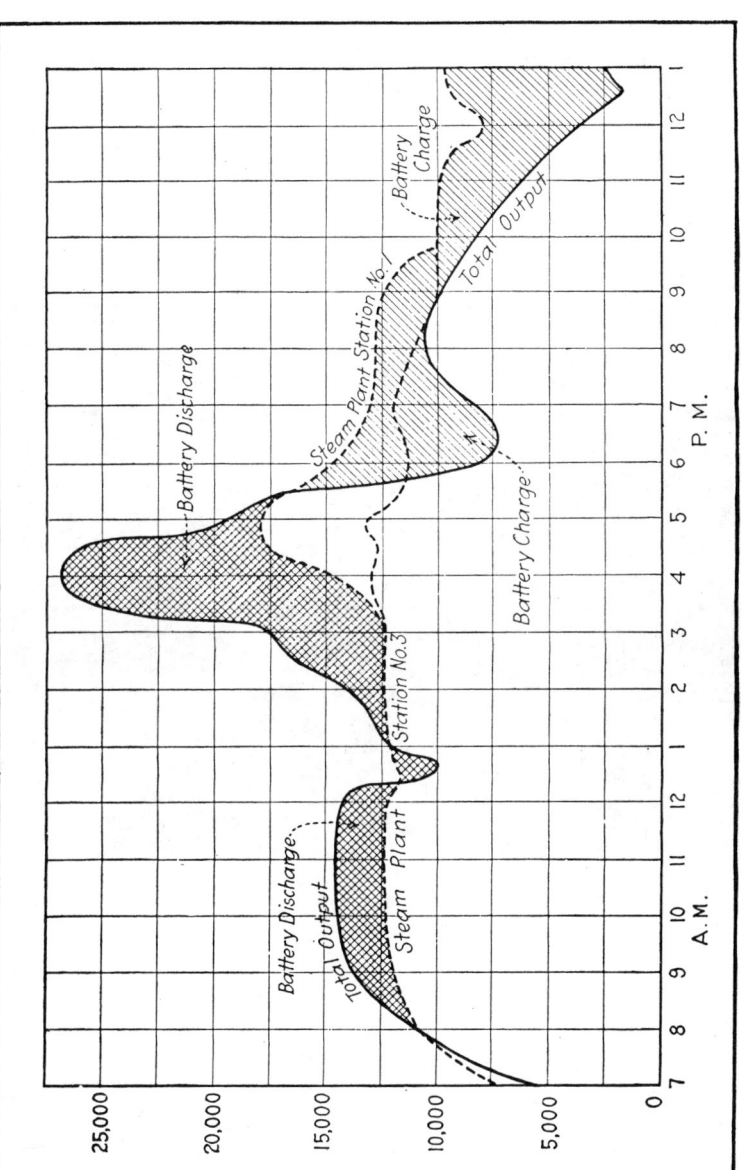

Fig. 86—How Exide Stand-by Battery Helped to Handle Unexpected Lighting Load.

The tank linings are of pure sheet lead, the lap seams being burned with the hydrogen flame. The upper edges of the lining extend over the edges of the tank and down outside clear of the tank faces. Drip points are provided, so spaced that they come between the tank supports. Good insulation of the cells is an essential. The Exide insulator consists of a glass body surrounded by an inverted petticoat, this forming an annular trough partly

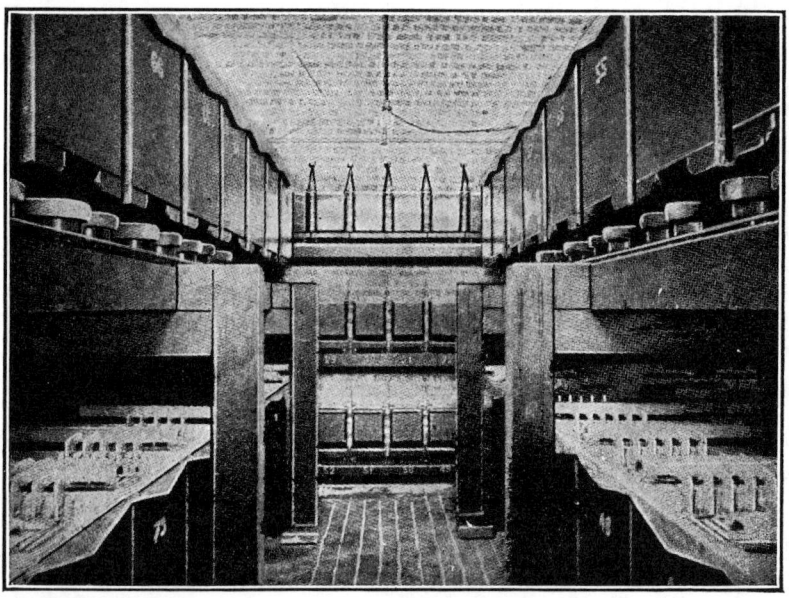

Fig. 87.—Battery-Room of Calumet River Draw-bridge of Calumet & Western Indiana Railroad.

filled with oil, the whole being covered with a lead cap extending down around the sides but out of contact therewith. Each oil insulator rests on a truncated cone or pedestal of earthenware or a composition not affected by acid. Each cell is covered with a plate of heavy glass, which serves to condense the acid spray. All conductors are of specially heavy section lead-coated copper bars, firmly bolted together. The plates of each cell, joined together

THE STAND-BY BATTERY

to form an element or plate group, are lead-burned to soft-lead busbars.

The total number of cells required is determined by the maximum busbar voltage permissible at the end of a high-discharge rate. For example, if a battery is installed of capacity enough to carry a maximum peak load for ten minutes a voltage of 100 each side of a three-wire system might be considered satisfactory. This would call for 75 cells each side. If at any time during the day it is necessary to float the battery at a voltage as low as 115 volts, the main battery would consist of 55 cells each side, leaving an auxiliary battery of 20 end-cells on each side. These end-cells are brought into action as needed by special switches.

In arranging a battery-room for such a large battery, special attention is given the floor construction, because drainage is important. Ventilation must be exceptionally good, and all exposed metal work should be protected by acid-proof paint. Concrete has been used for battery-room floors, and is satisfactory if care is taken to flush it with water frequently to wash away any electrolyte that may have accumulated. Hard-burned tile or vitrified brick is much more suitable. The floor should be laid on a concrete foundation, of sufficient strength to carry the weight. The slope should be such as to allow for positive drainage, and the floor covered with asphaltum felt. Spaces of $\frac{1}{4}$ inch are left between the bricks or tiles, these spaces being grouted with asphalt compound. An exhaust ventilating system is almost an essential if the battery is used much. The fan parts should be of bronze, and air from the room should be filtered through an air-box having perforated lead screen to eliminate acid spread. It is said that air in a battery-room should be changed completely four times an hour during the gassing period of the charge.

Storage Batteries for Draw-Bridge Operation.—An unfailing source of power is an absolute necessity where draw-bridges are operated by electric motors, as most of them are. This applies especially to railway bridges, where any failure of the power supply would seriously interrupt travel on either the waterway or railway. Vessels have been badly injured due to failure to open a draw. Bridges may be operated by separate power plants, consist-

Fig. 88.—Calumet & Western Indiana Railroad Bridge in Open and Closed Positions.

THE STAND-BY BATTERY

ing of dynamos driven by internal-combustion engines, or they may take their power from a trolley or lighting system. In either case, an auxiliary storage-battery installation insures absolutely reliable service, even in event of failure of the main source of current. This is of special value where a bridge is over a widely used stream. The main principles of operation and the requirements are practically the same as obtain when storage batteries are used for stand-by service. A typical installation, where the movable span of the bascule pattern is 186 feet long and weighs 1,100,100 pounds itself, in addition to a 3,000,000-pound concrete counterbalance, consists of two batteries, one to furnish power for the bridge motors, the other for signals and lighting. The larger battery consists of 120 type F11 "Chloride Accumulator" elements in lead-lined wooden tanks. Each element has a normal capacity of 400 ampere-hours at 240 volts. The general arrangement of the tanks in the battery-room is shown at Fig. 87. The views at Fig. 88 show the bridge in open and closed positions.

Edison Storage-Battery Mine Lamp.—The greatest danger in mines to-day is from the use of unprotected flames wherever the deadly fire damp is likely to be encountered. Artificial light is an essential which formerly could not be obtained with safety unless provided in such minute quantity as to seriously curtail production. The Davy safety oil lamp is well known and has been widely used, but electric lighting gives much superior results. Much ingenuity has been shown in trying to adapt portable electric lamps to this work, but the stumbling-block always has been the production of current for their operation. The only practical source of energy is naturally some sort of battery, and many attempts have been made to modify the old types of cells so that they would serve the purpose satisfactorily. The advocates of the primary battery soon found that the inherent defects of this type were greatly magnified when an endeavor was made to produce a portable form of small size and weight with sufficient capacity to keep an electric lamp burning any considerable time. Aside from this, the electrical energy is produced in a primary battery by the consumption of the zinc plates, so that there was constant expense and trouble for their renewal. The primary cell was early elimi-

188 STORAGE BATTERIES SIMPLIFIED

nated from serious consideration, and experiments made with secondary or storage batteries have until lately been far from successful. The Edison mine lamp, shown at Fig. 89, is considered a practical solution of this problem, because of the "meddle-proof" qualities of the Edison alkaline battery. The cells used are the same in principle as the nickel-iron alkaline batteries developed for other purposes, but are small and very light.

The cells fit snugly into a light case of *rust-proof steel,* which is primarily a box in which to carry the battery. There is no insulation between the cells, and the case and the contact springs on the cell poles hold the battery securely when the cover is in place.

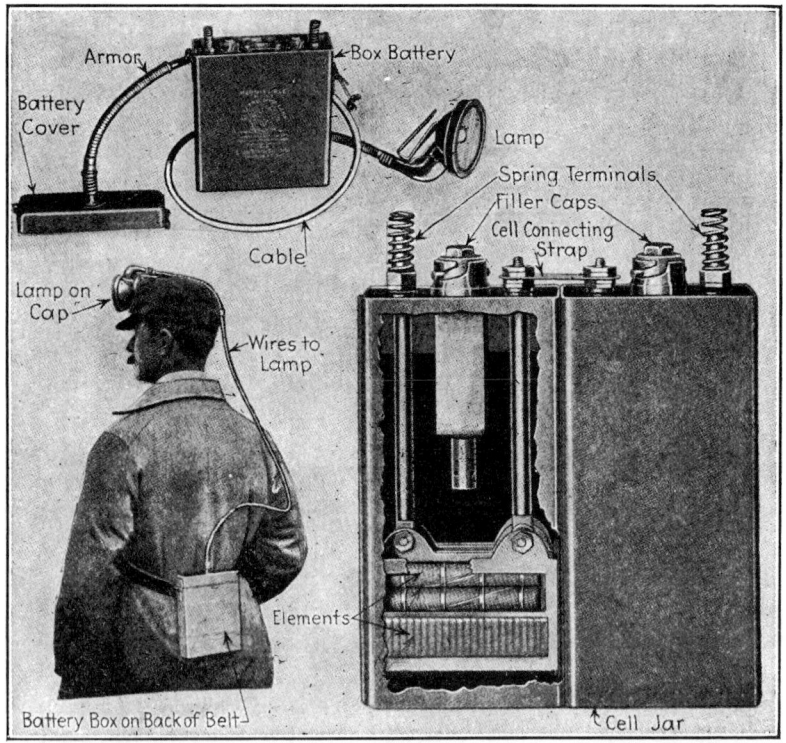

Fig. 89.—The Edison Storage Battery Mine Lamp Outfit and How It is Used.

EDISON MINE LAMP

The cover has a separable hinge, which permits its entire removal when open and facilitates charging the batteries in "banks." Covers and cases, equipped with self-contained locks, are interchangeable. The two cells are connected in series, the positive pole of one and the negative of the other being grounded to their containers, and the containers connected together. The free terminals carry the spiral contact springs, which press against nickeled-steel contact plates in the cover. The contact plates are insulated from the cover and receive the cable terminals. A twin-conductor, rubber-covered cable connects the battery to the cap lamp. At each end the cable is thoroughly armored, preventing injury from sharp bending. While lamp and reflector are being carried in the hand or at other times the armor takes up all the weight, so there is no possibility of strain coming upon the wires at the terminals. An ingenious arrangement permits the easy replacement of the cable should it be cut or otherwise injured in service. The cap lamp consists of a nickel-plated brass reflector provided with a hook to fit into the regulation miner's cap. A tungsten lamp is forced into a spring socket by means of a clip at its tip in such a way that if the lamp be broken the base is immediately disconnected and the lamp extinguished. This safety feature has been thoroughly tested by the Bureau of Mines and unqualifiedly approved under Schedule 6A.

CHAPTER VI

GLOSSARY OF STORAGE BATTERY TERMS

Acid.—As used in this book refers to sulphuric acid (H_2SO_4), the active component of the electrolyte.

Acidometer.—A hydrometer for testing specific gravity of acid, and specially graduated for that purpose.

Active Material.—The active portion of the battery plates; peroxide of lead on the positives and spongy metallic lead on the negatives of lead-plate types.

Alternating Current.—Electric current which does not flow in one direction only, like direct current, but rapidly reverses its direction or "alternates" in polarity so that it will not charge a battery.

Alloy.—A mixture of two or more metals produced by fusion, i. e., brass is an alloy of copper and zinc melted together; German silver is an alloy of copper, nickel and zinc.

Ampere.—The unit of measure of the rate of flow of electric current.

Ampere-Hour.—The unit of measure of the quantity of electric current; thus, 2 amperes flowing for $\frac{1}{2}$ hour equals 1 ampere-hour.

Anode.—The opposite plate to the cathode as the carbon plate of a primary battery. The anode is the terminal the current passes into from the solution.

Antimony.—A bright, bluish-white, brittle and easy-pulverized metal combined in small quantities with lead to form a harder alloy suitable for storage-battery plate grids.

Arc-Burning.—Making a joint by means of electric current, which melts the metal of the parts to be joined together.

Asphaltum.—A natural, tarry substance, not affected by acid, and also an electrical insulator, widely used as a basis for cell-sealing compounds.

Battery.—Any number of complete cells assembled in one set.

Battery Terminals.—Devices attached to the positive post of one end cell and the negative of the other, by means of which the battery is connected to the outer circuit.

Buckling.—Warping or bending of the battery plates.

Burning-Strip.—A convenient form of lead, in strips, for filling up the joint in making burned connections.

Busbar.—A main conductor, usually of heavy section, to which a number of circuit ends having the same characteristics are attached, to save wiring. All positive plates of a storage cell may be said to be attached to a busbar, the negative plates likewise. Instead of having a

GLOSSARY

separate conductor coming from each plate and joined to the outer circuit, the plates of a given polarity are grouped by attaching to a busbar.

Carbon.—One of the chemical elements. A black solid, that may exist as coal, charcoal or graphite, or, after it has been subjected to intense heat and pressure, as a white diamond. A conductor of electricity, having considerable resistance. It is used as a neutral plate in primary batteries, for lead-burning electrodes and for rheostat work. It is not affected by acid. Most of the carbon used in electrical work is manufactured and is not a natural product, as graphite.

Carboy.—A large glass bottle carried in a wooden case for easy handling. Used to hold acid, electrolyte or water.

Cathode.—The terminal of an electric circuit from which the current passes into the solution. The zinc plate of a primary battery is a cathode. The cathode is always the element of a battery most acted upon by the electrolyte.

Case.—The containing-box which holds the battery cells.

Cell.—The battery unit, consisting of an element complete with electrolyte, in its jar with cover.

Cell Connector.—The metal link which connects the positive post of one cell to the negative post of the adjoining cell.

Central Station.—A complete power plant equipped with large dynamos for supplying electric energy to an entire district.

Charge.—Passing direct current through a battery in the direction opposite to that of discharge, in order to put back the energy used on discharge.

Charge Rate.—The proper rate of current to use in charging a battery from an outside source. It is expressed in amperes, and varies for different-sized cells.

Chemical Change.—The uniting of certain primary or basic substances to form secondary ones, or the breaking apart of complex substances to determine their essential elements. Chemical combination is when elements form a new substance. For instance, hydrogen and oxygen gases combined in the proper proportions will form the liquid we know as water. Decomposition is the reverse of combining elements, Water may be decomposed and hydrogen and oxygen gas liberated by electrolysis.

Chemical Element.—These are basic substances, of which everything in the universe is composed. They may be solid, such as iron, zinc, lead or carbon; they may exist as a gas, such as hydrogen and oxygen, or as a liquid, such as bromine. Some elements combine rapidly with nearly all the others, and some cannot combine except with certain ones. Oxygen is the most active element, and will combine with many of the rest. There are about seventy-five of these elements, though practically everything on earth may be made by various combinations of less than twenty of them. The process of combining elements is known as "synthesis," that of separating them as "analysis." The elements which combine together the easiest are the hardest to separate.

Circuits.—An electrical circuit is said to be an open circuit when the current cannot flow, and a closed circuit if there is a continuous path for the electricity.

Circuit-Breaker.—An automatic, mechanical, electrically actuated device that takes the place of the fuse and performs the same function in an electric circuit.

Compound Winding.—A method of winding electric machines where both series and shunt windings are incorporated.

Conductor.—A pipe is a conductor of water. If two electrically charged bodies are connected by a piece of wood, glass, rubber, dry cloth, paper or similar materials, there will be no passage of electricity, but if a metal rod is substituted, a current will flow from the body of higher potential to the other. In this case the metal rod or wire is a conductor of electricity. All metals and substances such as acid, water and the various liquids (except oils) conduct electricity so well as to be termed "conductors," though it is harder for the electrical current to flow through some kinds of metal than it is for it to pass through others. Copper, aluminum and silver are very good electrical conductors, steel or iron come next in order, while some alloys, such as German silver, offer considerable resistance to the flow of current.

Contact Breaker.—A mechanical switch for closing and opening a circuit in rapid succession.

Controller.—A manually or automatically operated device for altering the current flow. Such a device may be interposed between a battery of an electric automobile and the driving motor to vary the speed and power of the latter.

Copper.—A reddish-brown metal widely used for electric wires and terminals because of its excellent conductivity. It is employed in many forms of primary battery as the plate of opposite polarity to the zinc element.

Corrosion.—The attack of metal parts by acid from the electrolyte; it is the result from lack of cleanliness.

Counter E.M.F.—A potential difference or voltage in a circuit opposed to the main voltage and resisting the flow of the latter. When charging a storage battery, the battery voltage is counter E.M.F. to that of the charging line.

Cover.—The rubber cover which closes each individual cell; it is sometimes flanged for sealing compound to insure an effective seal.

Current.—The passage of electricity through any piece of apparatus is termed a current. If the flowing of the electrical charge is continuous it is called a direct current. If the charges are not continuous

GLOSSARY

but flow always in the same direction it is termed a "pulsating" current. If an electrical charge flowing in one direction is followed by another charge flowing in the opposite direction, an "alternating" current is produced.

Cutout.—An electro-magnetic mechanism that automatically performs the same function of opening a closed circuit that a hand-operated switch does.

Discharge.—The flow of electric current from a battery through a circuit. The opposite of "charge."

Distilled Water.—The condensed water vapor or steam obtained by cooling vapors given off from boiling water. This will remove the impurities, such as salts, etc. These remain in the still as residue, only chemically pure water being vaporized.

Dynamo.—An electrical machine capable of producing current and distributing this current as desired, providing the current is sufficiently strong to overcome the resistance to its motion of the parts comprising the external circuit.

Electrolyte.—The fluid in a battery cell, consisting of specially pure sulphuric acid diluted with pure water in some cases and an alkaline solution in others.

Element.—One positive group and one negative group with separators, assembled together.

Electric Contact.—The joining of two conductors so a current can pass from one to the other.

Electrode.—The terminal of any open circuit.

Electrolysis.—The separation of a chemical compound into its constituents, by the action of an electric current. It cannot take place unless this compound is a conductor of electricity.

Electro-Magnet.—A bar of iron magnetized by passing a current of electricity through a coil of insulated wire wrapped around it. When the current is interrupted the iron bar or core piece ceases to possess magnetic qualities.

Electrical Distribution.—The action of an electrical machine in regulating the distribution of electricity may be considered to be the same as that of a pump which takes water from one tank and supplies it to another at a higher level. If for these reservoirs we consider bodies insulated from each other, we can, with an electrical generator, take electricity from one that has been overcharged and supply it to another which is undercharged.

Electricity.—A force that no one knows the exact nature of. To form some conception of this force, it is well to consider that we are able to place various bodies in different electrical relations. A stick of sealing wax or a hard rubber comb rubbed on a coat sleeve will attract bits of paper, feathers and other light objects. The sealing wax or rubber is said to be charged with electricity which has been produced by friction against the coat sleeve. Electricity may be produced by mechanical, chemical or thermal action.

Electrical Charge.—Any body charged with electricity may be considered one whose surface is supplied with either an overcharge or undercharge of electricity. The overcharged body always tends to discharge to the undercharged body in order to equalize a difference in pressure existing between them.

Filling Plug.—The plug which fits in and closes the orifice of the filling tube in the cell cover.

Flushing.—Replacing electrolyte in lead-plate cells with acid instead of distilled water.

Flooding.—Overflowing through the filling tube. With the usual vent this can occur only when a battery is charged with the filling plug out.

Freshening Charge.—A charge given to a battery which has been standing idle, to insure that it is in a fully charged condition.

Forming.—The process of making storage-battery plates from lead sheets by a series of charging and discharging operations.

Fuse Box or Fuse Block.—A non-conducting container for safety fuses, usually of porcelain, slate or marble.

Fuse.—An electrical safety valve to prevent an overload or passage of excessive amounts of current through a circuit. These are made of fusable lead alloy wire, which melts or "blows" if too much current is passed through it, thus breaking the circuit in which it is placed.

Gassing.—The bubbling of the electrolyte caused by the rising of gas set free toward the end of the charge.

Generator System.—An equipment including a generator for automatically recharging the battery, in contradistinction to a straight storage system, where the battery has to be removed to be recharged or coupled to an external current source.

Glass.—A fused mixture of silicate of various oxides, and a very good non-conductor of electricity if dry. Not affected chemically by most acids or alkali. May be made either opaque or transparent, depending upon coloring matter added. A very common, brittle substance, widely used for storage battery and primary cell jars, insulators and containing-vessels for all kinds of liquids.

Gravity.—A contraction of the term "specific gravity," which means the density compared to water as a standard.

Grid.—The metal framework of a plate supporting the active material, and provided with a lug for conducting the current and for attachment to the strap.

Group.—A set of plates, either positive or negative, joined to a strap. Groups do not include separators.

H_2O.—Chemical symbol for water.

GLOSSARY

Hard Rubber.—A rubber compound that has been hardened by heat treatment so it has greater stiffness than rubber in its natural form, and will keep its shape indefinitely after forming. This material is very brittle and not very strong. It is an excellent insulator of electricity, and as it is not affected by sulphuric acid it is widely used for cell jars.

Hold-Down Clips.—Brackets for the attachment of bolts for holding the battery securely in position on the car.

Horse-Power.—The accepted unit of mechanical work. The ability to move 550 pounds one foot in one second or 33,000 pounds one foot in one minute. An electrical horse-power is 746 watts.

H.P.—Abbreviation for horse-power.

Hydrogen.—One of the basic elements existing as a gas under natural conditions. It may be liquefied by the simultaneous application of great pressure and abstraction of heat. It is the lightest known substance. The chemical symbol is H.

Hydrogen Flame.—A very hot and clean flame of hydrogen gas and compressed air used for making burned connections.

Hydrogen Generator.—An apparatus for generating hydrogen gas for lead-burning.

Hydrometer.—An instrument for measuring the specific gravity of the electrolyte.

Hydrometer Syringe.—A glass barrel enclosing an hydrometer and provided with a rubber bulb for drawing up electrolyte.

Induction.—The creation of a current in a conductor not connected to a source of electricity by the juxtaposition of one that is carrying the current.

Induction-Magnetic.—The magnetization of any magnetic substance, such as iron or steel, by placing it in a magnetic field but not in actual contact with the energizing magnet.

Insulating Tape.—A textile fabric impregnated with insulating compound of an adhesive nature. Used to cover bare spots in insulated wires, re-enforce insulation, and for protecting joints where wires are joined together.

Insulating Varnish.—Shellac or sealing wax dissolved in alcohol, or gum copal dissolved in ether, may be used as a varnish for insulating purposes.

Insulator.—Materials such as wood, glass, rubber, etc., and air, conduct electricity so badly as to be termed insulators. What would normally be an insulator to a current of low potential may be ruptured by a current of higher potential or pressure which can break down the resistance.

Iron Oxide.—Commonly known as "rust." It is packed in steel pockets, which are assembled into negative plates of Edison Storage Battery. Expressed chemically as FeO.

Jar.—The hard rubber container holding the element and electrolyte.

KOH.—Chemical symbol for caustic potash or potassium hydrate.

Lead.—An abundant and widely used metal of bluish-white color, and one of the softest and heaviest of metals. It is not acted upon by sulphuric acid unless an electric current is passed through it. It forms the main part of most storage-battery plates, either as a metallic lead or as a lead oxide.

Lead-Burning.—Making a joint by melting together the metal of the parts to be joined.

Lead Oxide.—Material on plates when a cell is discharged according to some theories of storage-battery action. This is expressed chemically as PbO, differing from peroxide only because there is less oxygen combined with the lead.

Lead Peroxide.—The active material on positive plates of lead batteries after charging. Expressed chemically as PbO_2.

Lead Sulphate.—Material on storage-battery plates when cells are discharged, caused by absorption of sulphate from the electrolyte. Expressed chemically as $PbSO_4$.

Lime, Slaked.—First quicklime is obtained by burning limestone, chalk or marble in kilns and afterward removing its caustic properties by watering it and allowing it to remain in the air for a time. This is used in battery compartments of electric vehicles to neutralize spilled acid, as it is of an alkaline nature.

Litharge.—A yellow or reddish oxide of lead that is partially fused.

Local Action.—Wasteful oxidization of zinc in a primary battery when it is not in use, or abnormal sulphation of storage-battery plates due to impurities in the electrolyte.

Lug.—The extension from the top frame of each plate, connecting the plate to the strap or busbar.

Magnetism.—This is a property possessed by certain substances, and is manifested by the ability to attract and repel other materials susceptible to its effects. When this phenomena is manifested by a conductor or wire through which a current of electricity is flowing, it is termed "electro-magnetism." Magnetism and electricity are closely related, each being capable of producing the other.

Magnetic Substances.—Only certain substances show magnetic properties, these being iron, nickel, cobalt and their alloys. The earliest known substance possessing magnetic properties was a stone or iron ore first found in Asia Minor. It was called the "lodestone," or leading stone, because of its tendency, if arranged so it could move freely, of pointing one particular portion toward the north.

Magnetic Attraction.—If the north pole of one magnet is brought near the south pole of another, a strong attraction will exist between them, this depending upon the size of the magnets used and the air-gap separating the poles. Magnets will attract all magnetic substances.

Magnetic Repulsion.—If the south pole of one magnet is brought close to the end of the same polarity of the other there will be a pronounced repulsion of the forces. The like poles of magnets will repel each other because of the obvious impossibility of uniting two influences or forces of practically equal strength but flowing in opposite directions. The unlike poles of magnets attract each other because the force is flowing in the same direction.

Magnetic Flow.—The flow of magnetism is through the magnet from south to north, and the circuit is completed by the flow of magnetic influence through the air-gap or metal armature bridging it from the north to the south pole.

Maximum Gravity.—The highest specific gravity which the electrolyte will reach by continued charging, indicating that no acid remains with the plates.

Meters.—Most of the electrical measuring instruments depend upon the principle of electro-magnetism or induction. These measuring instruments are made in portable and switchboard types. The windings in an instrument designed to measure current quantity or amperage are usually of coarse wires, while the windings of an instrument to measure electro-motive force or voltage will be of finer wire. The gauge used to measure current quantity is called an ampere meter or ammeter, while that used to measure current pressure is a volt meter.

Mica.—An insulator of natural mineral derivation that will stand considerable heat. Not suited for use with high-potential currents, because it is apt to contain impurities of a metallic nature. Commonly known as "isinglass."

Motor.—A machine that is capable of delivering current in one direction when driven by mechanical power and which will produce mechanical energy if electric current is passed through the winding in a reverse direction.

Motor-Generator.—An electrical machine that may be used either as a current producer or for generating electricity if driven by mechanical means, or as a power producer if driven by electrical means.

Negative Pole.—The terminal of a current-generator to which the current flows after leaving the outer circuit.

Nickel.—A silver white malleable and ductile metal, that can be applied to others by thin surface coating through an electro-deposition or plating process.

Nickel-Hydrate.—A green powder used as the active material in the positive plates of the Edison storage battery.

Ohm.—The ohm is the unit by which resistance is judged. Everything has electrical resistance. Some elements have very little, such as a short length of good conductor; others have so much as to form a most effectual barrier to the passage of the current, these being commonly known as insulators.

Ohms Law.—The fundamental rules expressing the relation between voltage, amperage and resistance. It may be expressed thus: the current strength (C) is equal to the voltage or electro-motive force (E) divided by the resistance (R), or $C = E \div R$. Naturally, the voltage is equal to the current strength multiplied by the resistance, or $E = C \times R$. The resistance is equal to the voltage divided by the amperage, or $R = E \div C$.

Oil of Vitriol.—Commercial name for concentrated sulphuric acid (1.835 specific gravity). This is never used in a battery, and would quickly ruin it.

Oxidization.—The chemical combination of oxygen with any substance. Iron rust is ferrous oxide, and has been caused by oxidization.

Oxygen.—One of the most active of the elements that naturally exist as a gas, though it may be liquefied. Owing to its great affinity for various substances, it is not found free or uncombined. The chemical symbol is O.

Pb.—Chemical symbol for lead.

Paraffine Wax.—A white substance produced by distillation of crude petroleum, and one of the best insulators known.

Pickling.—The process of cleaning metal by dipping in an acid solution. This solution is known as a "pickle."

Plates.—Metallic grids supporting active material. They are alternately positive (brown) and negative (gray).

Polarity.—A difference in electrical condition. The positive terminal of a cell or battery, or the positive wire of a circuit, is said to have positive polarity; the negative, negative polarity.

Post.—The portion of the strap extending through the cell cover, by means of which connection is made to the adjoining cell or to the car circuit.

Positive Pole.—The terminal of a current-generator from which the current is intended to flow to the outer circuit.

Potassium Hydrate.—An alkaline substance combined with water to serve as the electrolyte in the Edison storage battery. Commonly known as caustic potash. Expressed chemically as KOH.

Potential or E.M.F. (Electro-Motive Force).—The greater the difference in the quantities of the electrical charge the greater the tendency to reach the state of equilibrium. This difference in electrical conditions, or amount of electrical charge, is termed "difference of potential," and high or low potential, or "electro-motive force," in any electrical system indicates a large or small difference of charge or electrical condition at different parts. This is measured in volts.

GLOSSARY

Rectifier.—Any device capable of changing alternating current to one having the properties of direct current.

Resistance.—Material (usually lamps or wire) of low conductivity inserted in a circuit to retard the flow of current. By varying the resistance, the amount of current can be regulated.

Resistance, External.—The resistance of those parts of the circuit outside of the dynamo or battery producing the current.

Resistance, Internal.—The resistance of the windings of a generator or that of the electrolyte and separators of a storage battery as distinguished from that of the parts comprising the outer circuit.

Resistance, Ohmic.—Resistance measured in ohms is a true resistance.

Return, or Ground.—The conductor which is supposed to carry the current to its starting-point after it has passed through parts of the outer circuit. In an automobile lighting, starting and ignition system the metallic chassis frame is often used as a "ground return" to the battery. In large power installations the earth is actually used as a return conductor.

Rheostat.—A device having coils of different resistance that can be brought into action progressively to control electric-current flow, as when charging batteries.

Rubber Sheets.—Thin, perforated hard-rubber sheets used in combination with the wood separators in some types of batteries. They are placed between the grooved side of the wooden separators and the positive plate.

Sealing Compound.—The acid-proof compound used to seal the cover to the jar.

Sealing Nut.—The notched round nut which screws on the post and clamps the cell cover in place in Exide batteries.

Sediment.—Active material which gradually falls from the plates and accumulates in the space below the plates provided for that purpose.

Series Winding.—A method of winding electric machines where the armature winding is in series with the field winding. All current produced in the armature coils must pass through the field coils as well before reaching the external circuit.

Separators.—Sheets of grooved wood, specially treated, inserted between the positive and negative plates to keep them out of contact.

Shellac.—A resinous, vegetable substance, soluble in alcohol, and having good insulating qualities.

Short Circuit.—A metallic connection between the positive and negative plates within a cell. The plates may be in actual contact or material may lodge and bridge across. If the separators are in good condition, a short circuit is unlikely to occur.

Shunt Winding.—A method of winding electric machines where the armature coils are in parallel with the field coils. Only a portion of the current produced in the armature passes through the field coil.

Spacers.—Wood strips used in some types to separate the cells in the case and divided to provide a space for the tie bolts.

Specific Gravity.—The density of the electrolyte compared to water as a standard. It indicates the strength and is measured by the hydrometer.

Starvation.—The result of giving insufficient charge in relation to the amount of discharge, resulting in poor service and injury to the battery.

Steel.—An alloy composed of iron and carbon in its simplest forms. It contains from .05 to 1.80 per cent. of carbon. It is a variety of iron that can be hardened and softened by heat treatment, as well as having all the properties of iron as regards malleability, etc. It forms an important part of Edison Storage Battery elements and container.

Strap.—The leaden casting or small busbar to which the plates of a group are joined.

Sulphated.—The condition of plates having an abnormal amount of lead sulphate caused by ''starvation,'' or by allowing battery to remain discharged for lengthy periods.

Switch.—A switch interposed in an electrical conductor will, when opened, leave an air-gap in this conductor that offers so much resistance to the flow of current that the electricity cannot pass. Closing the switch so that the continuity of the conductor is re-established will enable the current to flow.

Tie Bolts.—Bolts which, in some types, extend through the battery case between the cells and clamp the jars in position.

Time Cutout.—Cutout devices which automatically break the charging circuit of storage batteries when the current has passed through a sufficient time to insure proper charging. A time cutout is merely a switch operated by clock work.

Transformer.—A form of induction coil to ''step up'' or increase voltage or to ''step down'' or decrease voltage. As a rule, when the potential is increased the amperage or current is reduced and vice versa. An ignition coil transforms a current of low voltage and strong amperage to one of high potential and very low amperage. A transformer can also change high-voltage current to one of low potential and secure an increase in amperage.

Top Nut.—The hexagon nut which, in batteries with bolted connections, screws on the post and holds the connectors and sealing nut in place.

Vaseline.—One of the residues left after distilling off the lighter constituents of crude petroleum oil. Used as a coating for brass or copper terminal screws on storage batteries to prevent corrosion or

GLOSSARY

formation of verdigris by the chemical action of the electrolyte on the metal.

Vent.—Special fittings placed in the cover of sealed battery cells to allow passage of gas and prevent electrolyte from splashing out.

Voltage.—Electrical potential or pressure, of which the volt is the unit.

Watt.—A watt is a unit of quantity, or amount of electric energy, and corresponds to a current of one ampere at a pressure of one volt. Thus a watt is a volt-ampere-second, and 746 watts indicate an amount of electrical energy equal to one mechanical horse-power. A kilowatt is 1,000 watts.

Zinc.—A silvery white metal having a crystalline fracture and somewhat similar to lead in many respects, though not nearly so heavy. This material is widely used in primary batteries as the active plate, but is seldom made into storage-battery plates.

INDEX

A

	PAGE
Acid, Definition of	190
Acid Electrolyte Proportions	63
Acidometer	190
Acid, Properties of	63
Action of Depolarizer	15
Action of Primary Cell	13
Active Material	190
Advantages of "Iron Clad" Exide Plate	44
Alkaline Battery, Edison	26
Alkaline Electrolyte	66
Alloy	190
Alternating Current	190
Ampere	190
Ampere, Definition of	17
Ampere-Hour	190
Ampere-Hour Meter, Use of	151
Anode	190
Antimony	190
Antimony Alloys for Grids	38
Apparatus for Making Cadmium Tests	61
Arc-Burning	190
Arc-Burning Outfit	83
Asphaltum	190
Automobile Batteries, Ignition	124
Automobile Batteries, Power	136
Automobile Batteries, Starting	127

B

Batteries for Auto Work	23
Batteries for Draw-Bridge Operation	185
Batteries for Electric Automobiles	136
Batteries for Electric Launches	177
Batteries for Isolated Lighting	140
Batteries for Mine Lamp	187
Batteries for S t a n d - b y and Booster Work	180
Batteries for Starting Engines	129
Batteries for Street Railway Cars	167
Batteries for Submarine Boat	170
Batteries for Switch and Signal Work	179
Batteries for Yacht Lighting	176
Batteries of Locomotives	162
Battery	190
Battery Capacity, Loss of	56
Battery Capacity, Rules Governing	25
Battery-Charging Apparatus	99
Battery-Charging Methods	91
Battery Charging, Direct Current for	94
Battery Charging in Garages	93
Battery Defects Summarized	90
Battery for Train Lighting	157
Battery Repair Tools	74
Battery, Secondary	18
Battery Shifts Gears	134
Battery Terminals	190
Buckling	190
Burning-Strip	190
Busbar	190

C

Cadmium Readings, Value of	60
Carbon	191
Carboy	191

INDEX

	PAGE
Carboy Stand	101
Care of Lead Batteries	81
Case	191
Cathode	191
Cause of Sulphation	58
Cell	191
Cell Cleanliness, Effect of Sediment	56
Cell Connector	191
Central Station	191
Charge	191
Charge Rate	191
Charging Resistance, Lamps for	109
Charging Vehicle Batteries	117
Chart of Electrolyte Mixtures	64
Chemical Action, Edison Battery	27
Chemical Action, Reversible	18
Chemical Change	191
Chemical Change in Lead Plates	21
Chemical Element	191
Chemical Production of Current	13
Chloride Plates	40
Circuit Breaker	192
Circuits	192
Combination Batteries	38
Commercial Type Edison Cell	46
Compound Winding	192
Conductor	192
Contact Breaker	192
Controller	192
Copper	192
Copper-Lead Storage Cell	33
Corrosion	192
Counter E.M.F.	192
Couple Type Storage Battery	29
Cover	192
Cure of Sulphation	58
Current	192
Current by Chemical Action	13
Cut-Out	193

D

	PAGE
Dangers of Flushing	57
Defects in Storage Battery	53
Delco-Lite Outfit	155
Depolarizer, Use of	15
Directions for Lead-Burning	87
Discharge	193
Dismantling Exide Cells	69
Dismantling Gould Cells	76
Distilled Water	193
Double Seal Exide Cover	69
Draw-Bridge Batteries	185
Dry Battery	15
Dry-Cell Connections	16
Dry-Cell Depolarizer	15
Dry-Cell Disadvantages	17
Dynamo	193
Dynamo for Train Lighting	158

E

Early Planté Cell	21
Edison Cell Electrolyte	66
Edison Negative Plate	48
Edison Plates	27
Edison Positive Plate	47
Edison Steel Cell Jar	50
Edison Storage Battery	26
Edison Storage Battery, Commercial Type	46
Electric Contact	193
Electric Launch Batteries	177
Electric Vehicle Batteries	137
Electrical Charge	194
Electrical Distribution	193
Electricity	193
Electricity by Chemical Action	13
Electrode	193
Electro-Magnet	193
Electrolysis	193
Electrolyte	193
Electrolyte, for Edison Cells	66
Electrolyte, Making	63
Electrolytic Rectifier	105

INDEX

	PAGE
Element	193
Element, Replacing	77
Engine-Starting Batteries	129
Essentials in Battery Care	53
Exide Batteries, Repair of	68
Exide Double Seal	69
Exide Single Seal	69

F

Faure, or Pasted Plates	23
Faure Plate Grid	38
Faure Type Plates	38
Figuring Mine Locomotive Capacity	163
Filling Plug	194
Flooding	194
Flushing	194
Flushing, Dangers of	57
Formed Plates, Planté	22
Forming	194
Freshening Charge	194
Fuse	194
Fuse Box	194

G

Gas Generator, Action of	85
Gas Generator, Hydrogen	85
Gear Shifting by Battery	134
Gassing	194
Generator System	194
Glass	194
Gould Battery Repairs	76
Gould Plates, Manufacture of	35
Gravity	194
Grid	194
Grids, Alloys for	38
Group	194

H

Hard Rubber	195
Hold-Down Clips	195
Horse-power	195
How to Find Cells in Poor Condition	59

	PAGE
How to Use Lead-Burning Outfits	87
Hydrogen	195
Hydrogen Flame	195
Hydrogen Gas-Burning Outfit	84
Hydrogen Gas, Generation of	84
Hydrogen Generator	195
Hydrometer	195
Hydrometer and Its Use	101
Hydrometer Syringe	195

I

Ignition Batteries	124
Ignition Systems	124
Illuminating Gas—Oxygen Outfit	86
Induction	195
Induction, Magnetic	195
Insulating Tape	195
Insulating Varnish	195
Insulator	195
Iron Clad Exide Battery	42
Iron Oxide	195
Isolated Lighting Plants	140

J

Jar of Edison Battery	50
Joining Dry Cells	16

L

Lamp-Bank Resistance	109
Lead	196
Lead Batteries, General Care of	81
Lead-Burning	196
Lead-Burning, Instructions for	87
Lead-Burning Outfits	81
Lead Oxide	196
Lead Peroxide	196
Lead Plates, Chemical Change in	21
Lead Sulphate	196
Lighting Battery Installation	148

INDEX

	PAGE
Lighting Battery, Operation of	147
Lighting Battery, Rules for Selection	145
Lime, Slaked	196
Litharge	196
Local Action	196
Locomotive Power Batteries	162
Loss of Battery Capacity	56
Lug	196

M

Magnetic Attraction	196
Magnetic Flow	197
Magnetic Repulsion	197
Magnetic Substances	196
Magnetism	196
Making Electrolyte	63
Making Planté Plates	35
Marine Applications, Miscellaneous	176
Maximum Gravity	197
Mechanical Forming, Planté Plates	37
Mercury Arc Rectifier	106
Meters	197
Mica	197
Mine-Lamp Battery	187
Mine-Locomotive Figuring	163
Miscellaneous Marine Applications	176
Motor	197
Motor Generator	197
Multiple Connection	16

N

Negative Plate, Edison	48
Negative Pole	197
Nickel	197
Nickel Hydrate	197

O

Ohm	197
Ohm's Law	198

	PAGE
Oil of Vitriol	198
Open Cell Type, Storage Cell	29
Oxidization	198
Oxygen	198

P

Paraffine Wax	198
Pasted Plates, Faure	23
Pickling	198
Planté, or Formed Plates	22
Planté Process Plates, Making	35
Plate-Burning Rack	88
Plate Deterioration, Cause of	60
Plates	198
Plates of Edison Cell	27
Polarity	198
Positive Plates, Edison	47
Positive Pole	198
Post	198
Potassium Hydrate	198
Potential, or E.M.F.	198
Primary Cell, Dry Type	15
Primary Cell, Simple	13
Principles of Storage Battery	18

R

Railway Switch and Signal Service	179
Rectifier	199
Rectifier, Electrolytic	105
Rectifier, Mercury-Arc	106
Rectifier, Wagner	108
Repairing Exide Batteries	68
Repairing Gould Sealed Types	76
Repairing Willard Battery	78
Replacing Gould Element	77
Resistance	199
Resistance, External	199
Resistance, Internal	199
Resistance, Ohmic	199
Resistance, Water	110
Return, or Ground	199
Reversible Chemical Action	18
Rheostat	199

INDEX

	PAGE
Rubber Sheets	199
Rules Governing Battery Capacity	25

S

Sealing Compound	199
Sealing Exide Cells	72
Sealing Nut	199
Sealed Type Storage Battery	29
Sediment	199
Sediment in Cells	56
Separator Functions	50
Separators	199
Separator Types	50
Series Connection	16
Series Multiple Connection	16
Series Winding	199
Shellac	199
Short Circuit	199
Shunt Winding	199
Simple Primary Cell	13
Simple Storage Battery	18
Single-Seal Exide Cover	69
Slow-Charge Cure for Sulphate	58
Spacers	200
Specific Gravity	200
Specific Gravity, Baumé Conversion	55
Stand-by Batteries	180
Stand for Carboys	101
Starting and Lighting Batteries	127
Starting-Battery Location	130
Starvation	200
Steel	200
Storage Batteries, Types of	29
Storage Batteries Using Other than Lead Plates	30
Storage Batteries, Winter Care of	112
Storage Battery Defects	53
Storage Battery for Ignition	124
Storage Battery History	19

	PAGE
Storage Battery Locomotives	162
Storage Battery, Principles of	18
Straightening Buckled Plates	70
Strap	200
Street Railway Car Batteries	167
Submarine Boat Batteries	170
Sulphate	200
Sulphation, Cause and Cure	58
Switch	200
Switchboard for Lighting	149

T

Taking Cadmium Reading	60
Tank Type Storage Battery	29
Tie Bolts	200
Time Cut-Out	200
Tools for Battery Repairs	74
Top Nut	200
Train-Lighting Battery	157
Train-Lighting Dynamo	158
Transformer	200
Types of Storage Batteries	29

U

Use of Hydrometer	101

V

Vaseline	200
Vehicle Batteries, Charging	117
Vent	201
Vibrator Rectifier, Westinghouse	96
Vitriol	198
Voltage	201
Volt, Definition of	17

W

Wagner Rectifier	108
Water Resistance	110
Water Used in Electrolyte	65
Watt	201

	PAGE
Watt, Definition of	17
Westinghouse Vibrator Rectifier	96
Willard Battery Repairs	78
Winter Care of Batteries	112
Wood Separators, Care of	51
Wood Separators, Faults of	70

X

	PAGE
Yacht-Lighting Batteries	176

Z

Zinc	201
Zinc Plates for Storage Batteries	33